The Blue Planet

Also by Louise B. Young

POWER OVER PEOPLE

EARTH'S AURA

The Blue Planet

LOUISE B. YOUNG

with Illustrations by Jennifer Dewey

Little, Brown and Company — Boston – Toronto

Second Printing

The author gratefully acknowledges the following for permission to reprint previously copyrighted material:

Selections abridged from *In the Deserts of This Earth* by Uwe George, translated
from the German by Richard and Clara Winston. Copyright © 1976 by Hoffmann
und Campe Verlag; English translation copyright © 1977 by Harcourt Brace
Jovanovich, Inc. Reprinted by permission of Harcourt Brace Jovanovich, Inc.

"Pliny the Younger's Letter to Cornelius Tacitus," from *Pliny Letters and
Panegyricus* Volume I, translated from the Latin by Betty Radice. Copyright
© 1969 by the President and Fellows of Harvard College. Reprinted by permission of Harvard University Press (Loeb Classical Library).

Excerpts from *The Travels of Marco Polo*, Revised from Marsden's Translation
and edited with an Introduction by Manuel Komroff. Copyright 1926 by Boni &
Liveright, Inc.; copyright renewed 1953 by Manuel Komroff. Copyright 1933 by
Horace Liveright, Inc. Reprinted by permission of Liveright Publishing Corporation.

"This Quiet Dust," by John Hall Wheelock, copyright 1919 by Charles Scribner's
Sons; copyright renewed 1947 by John Hall Wheelock; in *The Bright Doom*,
copyright 1927 by Charles Scribner's Sons; copyright renewed © 1955 by John
Hall Wheelock. Reprinted by permission of Charles Scribner's Sons.

LIBRARY OF CONGRESS CATALOGING IN PUBLICATION DATA

Young, Louise B.
  The blue planet.

  Bibliography: p.
  1. Earth sciences. I. Title.
QE26.2.Y68 1983   551   83-9425
ISBN 0-316-97707-1

MV

Designed by Susan Windheim
Published simultaneously in Canada
by Little, Brown & Company (Canada) Limited

PRINTED IN THE UNITED STATES OF AMERICA

To my grandchildren
with the hope that the earth they know throughout their
lives will be a place where violets still grow unmolested on
the forest floor, where morpho butterflies illuminate the
jungle shadows, and honey creepers still sip nectar from
the cup of gold.

Foreword

This book is written for all men and women who are curious about our home in space and its journey through cosmic time. It is planned for amateurs in the literal sense — *amatores* in Latin meaning lovers. I have written for those who love the earth, who find joy in sensing its moods, speculating on its past, its meaning, and its destiny. Many such amateurs may never before have opened a book about earth sciences, impeded by the belief that anything scientific would be too difficult to understand. These are the readers I especially want to reach, to open these doors that have been closed to them and introduce them to the most exciting and most rapidly changing field of knowledge today. Therefore I have cast my book in terms that require no previous scientific exposure. Amateurs also include a special type of men and women who have made a study of some scientific specialty their life work but who have retained their sense of wonder about all nature and enjoy looking beyond the confines of their own fields. They are natural philosophers by bent and avocation.

In *The Blue Planet* I have not attempted to present a solid and balanced coverage of all aspects of the earth sciences. Rather, I have concentrated on those aspects that seem to me particularly significant, mysterious, or beautiful. I have tried to suggest the unique personality of our planet, the characteristics that have been molded by all the experiences through which it has passed. Intimations of the future are there, too; going beyond the conventional boundaries of science to include thoughts about the nature of the universe and our own role in the cosmic process.

In writing this book I have drawn my material from many different fields of knowledge and have needed the advice of experts in at least half a dozen specialties. I have been fortunate in having the cooperation of many distinguished scientists. In some cases they read chapters dealing with their fields of expertise and made suggestions that I incorporated into the text. For such very constructive help I am indebted to Rainer Zangerl, Professor Emeritus, Field Museum of Natural History in Chicago; Thomas J. M. Schopf, Robert C. Aller, Paul B. Moore, and J. John Sepkowski, Jr., all professors of geophysical sciences, University of Chicago; and Linda Wilson, Curator of Education, Shedd Aquarium in Chicago.

Many other scientists have also been generous with their time, answering my questions, describing their fields of special interest, and providing valuable source materials. Among these, I especially want to thank James R. Heirtzler, Woods Hole Oceanographic Institution; Alfred M. Ziegler, Lawrence Grossman, and Frank M. Richter, professors of geophysical sciences, University of Chicago; R. W. Grigg, Hawaii Institute of Marine Biology; and Seiya Uyeda, professor of geophysics, University of Tokyo.

In spite of the expert assistance I have received, final responsibility for the accuracy and validity of the text resides with me. The interpretation of the facts, and in many cases the speculations about their meaning when set in a broader context, are mine alone.

LBY
Winnetka, Illinois
September 1983

Contents

List of Maps

The Blue Planet

Introduction

To see the world in a grain of sand
And a heaven in a wild flower;
Hold infinity in the palm of your hand
And eternity in an hour.

— WILLIAM BLAKE,
Auguries of Innocence

Viewed from afar, the earth is a luminous blue bubble wrapped in a swirl of filmy clouds and ethereal vapor. It is the only blue planet in the solar system. The surfaces of Mercury, Mars, and Pluto appear dark gray or rusty red because these bodies do not have enough atmosphere to soften and diffuse the harsh sunlight that falls upon them. Venus is enveloped in light yellow clouds; Jupiter is variegated, a swirling mass of red, green, orange, and brown. Neptune and Uranus are a hazy green and Saturn is pale gold, framed in its shining aureole of ice.

We can appreciate the special qualities of our own planet now that we have planted our footprints on the dusty face of the moon, have lowered our instruments into the hot inferno of Venus, have photographed the bare rocks of Mars, and stared into the turbulent whirlwind of the great red spot on Jupiter. We have measured the jagged forms of asteroids that swing in eccentric orbits around our star and watched the plumes of volcanic matter streaming upward from the ruptured surface of Io. These are awe-inspiring but forbidding landscapes. The more we explore the black expanses of space, the more we recognize that our own earth is an exceptional planet — the Garden of Eden of the solar system. There is no other piece of matter within a hundred billion miles that is so richly endowed with variety and beauty, dressed in its magnificent, ever-changing panoply of life.

Close up, the earth has a fine-grained, delicate perfection in the shape and texture of little things; the downy wings of the dandelion seed, the fluid line of the swallow's flight, the iridescent feathers on the hummingbird's throat. The commonplace inanimate objects have their own architectural beauty, although long familiarity has dulled our appreciation of them. We crush the surface of the earth beneath our feet and describe it in derogatory terms, calling it the "dirt," the "soil," the "dust." And we hasten to wash it off our hands. But when we look closely at these little pieces of the earth, we find that they are collages of exquisitely wrought miniature crystals with shining surfaces like the facets of a jewel. Miracles of artistic design and fascinating clues to the life story of the earth can be discovered in a pebble or a grain of sand.

It seems very likely that the crust of the earth resembles closely that of other planets — Mars, Mercury, possibly Venus. The rocks brought back from the moon look almost like stones picked up in our own backyard. But no water trickles over the rough surfaces of the moon; no atmosphere wraps it in a soft, protective veil. The moon — and probably Mercury, too — are harsh and silent worlds; no life can exist there. The surface of Venus is almost 900° F, hot enough to melt lead.

A few decades ago the imagination of man endowed the planet Mars with intelligent beings: builders of canals and irrigation systems and spaceships to invade the earth. Now the truth is known. If any life exists there, it is microscopic and quite different from any life forms on earth. The discovery has made us lonelier and wiser; suddenly we are aware that the earth is a very special place. In some ways it may even be unique in the universe.

Life has probably evolved in many other corners of the cosmos, on planets that have just the right degree of warmth, the right elements in their atmosphere and surface to nourish living things. So dense forests may clothe unknown continents, and oceans thick with aquatic life may lap on strange, shaly shores light-years away across the galaxy. But still the chances of duplication of life on earth are remote indeed. There may be no other planet that has monarch butterflies or nightingales or violet-colored sea fans that sway in warm ocean currents. Blooming plants did not evolve on earth until it was almost four and a half billion years old. Perhaps there is no other planet that has flowers.

In the Himalayas of Nepal and Tibet a Buddhist chant is inscribed on prayer stones, prayer wheels, prayer flags, and even on the living rocks: OM MANI PADME HUM. Carried silently on the wind (so the Buddhists believe), the incantation spirals upward into the pure bright air above the radiant peaks of the highest mountains in the world. Translated literally, the chant goes like this: "Om! The jewel in the heart of the lotus. Hum!" Interpreted freely, it means, "the Truth within the unfolding world of phenomena, all change and all becoming."

We who inhabit the planet earth are still far from knowing the Truth about our home in space, but every day we come closer to an understanding of the world of phenomena. The leaves of the lotus are unfolding. In recent years we have discovered many strange facts. We have totally revised our view of change and becoming.

Less than a century ago the earth was believed to be the very embodiment of stability, the unchanging background against which the drama of movement and growth and life was enacted. Men spoke confidently of the "solid earth" and the "everlasting hills." Now we know that nothing is static — not the earth itself nor any part of it. If we could watch a time-lapse movie of the planet's history, we would see an amazing drama of change and development: mountains being created and destroyed, seafloor ejected along the ocean ridges and consumed again at the trenches, canyons being carved by turbulent rivers, new continents split from old ones and set adrift to wander around the planet. This information has caused a revolution in our understanding of the earth, and the revolution is still in progress. Those of us who would know the planet we live upon must look at it anew with unjaded eyes like those of a child.

So we will go on a journey of discovery. We will travel far and wide across the earth's varied surface — look down into the gem pits of Sri Lanka, the gold mines of Cripple Creek, and the dark, glowing crater of Mount Etna. We will watch while scientists explore the crevasses beneath the sea, measure the strange stripes they have found on the ocean floor, and record the earth movements that make its surface tremble. We will listen while philosophers ponder the meaning of these things. But above all, we will harken to the voice of the poet who can see the world in a grain of sand and find the jewel in the heart of the lotus.

1

Profile of the Planet

> Mountains are immense things. . . . In order to
> exhaust their marvels and grasp the work of the
> Creator, one must love their spirit, study their
> essential features, wander about them widely,
> satiate the eyes and store up the impressions in the heart.
>
> — Kuo Hsi, *The Great Message of*
> *Forest and Streams*, 13th century

In the Forbidden City in Peking a spectacular piece of jade sculpture is displayed. A work dating from the Ming Dynasty (about A.D. 1400), it is a single monolith almost four feet high and five or six feet wide at the base, intricately carved into a soaring mountain landscape with steep precipices and deep ravines and cascading waterfalls. Clusters of tiny houses, elegant pavilions, and the stepped profiles of Chinese pagodas are nestled in small, protected valleys. Streams of miniature people, goats, and horses toil up the roads that zigzag across the mountain face. This ancient work of art captures the essence of the awe-inspiring landscape of our planet — its finely chiseled detail, its fascinating diversity.

The Chinese people were the first to depict in artistic form the majestic beauty of the face of the earth. As early as A.D. 1000 they were drawing in ink and color on silk or paper scrolls the dramatic natural features of their landscape: the very mountainness of mountains, the cool fluidity of waterfalls. In fact, *landscape* in Chinese means *mountains* and *water*.

Throughout the next thousand years this almost religious reverence for nature and a spiritual need for daily contact with it pervaded Chinese culture. Gardens were designed to capture in a tiny space the infinite landscape. Interesting weather-beaten rocks symbolizing the mountains were set on pedestals like pieces of sculpture. Fountains and reflecting pools added the essential ingredient of water.

Oddly enough, the Western peoples did not share this admiration for the wild and rugged scenes of nature until very recently. As late as the end of the seventeenth century Europeans were expressing aesthetic revulsion at the disorderliness of the earth's profile. Scientists such as Robert Hooke and Thomas Burnet were expounding to learned societies that the mountains were caused by crustal failure and collapse. Originally, they said, the planet had been a smooth and perfect sphere but now it is a mere "ruin" or "broken globe." In the mid–eighteenth century Casanova reported that when he traveled through the Alps he drew the blinds in his coach to spare himself the view of those vile excrescences of nature, the deformed mountains.

It was not until the late eighteenth century that appreciation of nature began to find expression in Western culture under the influence of men like Jean-Jacques Rousseau. In 1834 young Charles Darwin, exploring the Cordillera of the high Andes, exclaimed over the beauty of this mountain landscape. "It is sublime," he said, "like hearing a chorus of the *Messiah* in full orchestra."

Now most of the civilized world shares this enthusiasm. Unlike Casanova we do not pull down the blinds. We seek out the dramatic landscapes of snowcapped peaks and upland meadows drenched with sunlight. As Kuo Hsi advised, we wander about them widely, satiate the eyes, and store up the impressions in the heart.

The most treasured impression of mountain scenery stored in my memory is the view of the Annapurna range illuminated by the first rays of the rising sun. I had spent the night in Pokhare on the banks of Lake Phewa in Nepal. This valley is only 2,500 feet above sea level, verdant with banana trees and rice paddies and date palms. Behind this sea of green, like a dramatic backdrop, the Himalayas rise to awesome heights: 27,000, 28,000 feet. The clear blue water of the lake reflects their shining summits.

I slept only fitfully that night, disturbed by the pounding of rain on the roof and the sound of rising winds, heralding a storm. But when I woke half an hour before dawn I found a gibbous moon sailing free in a clear and almost cloudless heaven. The forms of the great mountains rose ghostly pale against a dove-gray sky, which slowly brightened. The stars faded and disappeared one by one. But yet all the land was wrapped in shadow when suddenly the top peak of Dhaulaghiri turned from silver to soft pink and then to gold. As I watched, the crest of Annapurna and the jagged summit of Machla Puchhare caught fire. Wispy clouds formed, floated across the scene, changed to apricot and flame. Gradually, peak by peak, the whole range was suffused with light as the rising sun sought out and gilded the domes and spires of this cathedral of the earth.

The village of Pokhare, still immersed in the dim religious light of predawn, began to stir. A cock crowed, shattering the stillness, and was answered by another. But the people rose and went about their tasks with muted sounds, reverently, as though awed by the splendor overhead. Women in bright saris, carrying baskets on their backs, moved in a stately procession down the roads to the rice fields. Across the waters of the lake boys propelled dugout canoes, chanting as they paddled. On a terrace by the lake a young girl trimmed the grass with a curved knife and sang softly in time with the rhythm of her movements. Far above the housetops a flight of egrets, white as angel wings, passed from the shadow of the range and were instantly transformed to flamingo color as they caught the light of the eastern sky — the glowing rose windows of returning day.

⌘

At higher altitudes mountain scenery takes on a clarity unknown at lower levels. On a fine day the sky is piercingly blue; the air is clean and cool. The haze and smog that are part of our normal daily experience are swept away as though a veil had been lifted, disclosing the true form and radiance of things. Euphoria may be experienced in these settings as Peter Matthiessen vividly described, recalling a scene on his trek to the Crystal Mountain in Tibet:

Left alone, I am overtaken by that northern void — no wind, no cloud, no track, no bird, only the crystal crescents between peaks, the ringing monuments of rock that, freed from the talons of ice and snow, thrust an implacable *being* into the blue. In the early light, and rock shadows on the snow are sharp; in the tension between light and dark is the power of the universe. . . . The sun is roaring, it fills to bursting each crystal of snow. I flush with feeling, moved beyond my comprehension, and . . . the warm tears freeze upon my face. These rocks and mountains, all this matter, the snow itself, the air — the earth is ringing. All is moving, full of power, full of light.

When human beings climb to even greater heights — above 14,000 or 15,000 feet — they encounter more and more arduous conditions. They begin to suffer headaches and nausea from insufficient oxygen. Intense cold, biting winds up to 70 miles an hour make the scaling of the world's highest peaks an extreme test of endurance, and this in itself becomes a tantalizing challenge. After the trail has been broken by the explorers, scientists are usually not far behind. They follow with their battery of instruments and their insatiable curiosity about the nature of the world we inhabit.

In 1963 the National Geographical Society and the National Science Foundation financed an expedition to the top of Mount Everest. Teams of scientists went along. They carried a delicate gravity meter by hand from the States; brought a seismograph to measure the thickness of the glaciers. Weather information was gathered regularly. Glaciologists studied the ablation (the surface wasting) and the movement of the ice. Geologists collected samples of rock, sometimes under the most perilous conditions, hanging on wire ladders eighty feet down in crevasses. Doctors, sociologists, and psychologists observed the physical and mental changes brought on by the extreme physical stresses of the high mountain environment.

On May 1 James Whittaker and the Sherpa mountaineer Nawang Gombu made the first attempt at the summit. It was a very unfavorable day with the temperature at 20° below zero F. Fierce winds sweeping across the mountain tore at the heavy packs of the climbers and nearly swept them off the narrow ridge which they followed up the southeast slope. For almost seven hours they fought their way upward step by step, zigzagging back and forth between layers of rock and treacherous snow cornices. At last they stood at the very top of the highest peak in the world. At their feet lay a great immensity of cloud and earth and air. Four countries were spread out there: the jagged peaks of Nepal, the rolling hills of Tibet, the great massif of Sikkim, the hot plains of India almost concealed beneath a bed of white foam. The near distance with its pinnacles of ice and stone was laser-sharp but far out toward the perfect circle of the horizon the forms were softened with a faint purple haze. Seen through two hundred miles of atmosphere, the meeting place of land and sky was blended in a froth of filmy white and violet blue like a view of the earth from space.

Several weeks later two other teams from the expedition followed Whittaker and Gombu to the top. The most interesting route was taken by the third team, Thomas Hornbein and William Unsoeld. They made the first successful ascent of the West Ridge, went over the top, and descended the Southeast Ridge, achieving the first traverse of any major Himalayan peak. The night before the climb they slept at a camp perched at 27,250 feet on the west face of the mountain. Early in the morning of May 22 they started up the treacherous West Ridge, which had never been scaled before. The snow was very hard and granular, so slippery that they were forced to cut steps for a distance of several hundred feet. Then they encountered a steep slab of yellow rock that was too crumbly to offer any firm foothold. Hornbein, who was in the lead, tried one attack after another. Finally he drove a large piton deep into the soft, "rotten" rock. The piton held, but Hornbein, exhausted by his efforts, was unable to negotiate the outcrop wall. Unsoeld took off his mittens and, using the piton for leverage, worked his way slowly over the rock face. Hornbein followed, using the same technique, and in that way they both finally surmounted the treacherous wall. Above them on the mountainside they could see patches of gray stone protruding through the snow. They knew that this surface would be firmer and easier to climb. But it was after six o'clock when Hornbein and Unsoeld finally reached the summit. The sun, low in the sky, cast the long black shadow of Mount Everest onto the slopes of Makalu. Almost the entire descent of the mountain was made in total darkness.

❧

The outcrop of "rotten" yellow rock that gave Hornbein and Unsoeld so much trouble is one of the most interesting features of this astonishing mountain. It is a soft limestone layer that lies in undulating curves across the west face of Mount Everest near the very peak and also in prominent bands along the saw-tooth ridge connecting the nearby peaks of Lhotse and Nuptse. Examination of samples of this limestone reveals that it was laid down in warm shallow seas 300 million years ago. Now it is covered deep in snow five miles above the sea.

In the Himalayas, as in many of the other mountain ranges on earth, distinct layers of rock, interbedded like an elaborate multidecker sandwich, are severely bent and twisted, bearing witness to the fact that the earth in these places has been subjected to extreme pressure. There is also abundant evidence that many of the strata were formed at or below sea level.

Samples of the Yellow Band brought back by the American expedition and an earlier Swiss expedition contained fragments of crinoids, plantlike animals with stems and root systems that grew in great abundance in the ancient seas. These remains are the highest fossils that have been found anywhere in the world. In somewhat lower and more accessible mountain regions — but still miles above sea level — many remains have been discovered of creatures that lived millions of years ago in the sea. When Charles Darwin explored the Andes, he was astonished to come upon a bed of fossil seashells at 14,000 feet and then a little farther down the mountainside, a small forest of white petrified pine trees with marine rock deposits around them.

Fossils are objects of striking form and occasionally great beauty, so it is not surprising that they have been noticed and commented upon since very early times. Throughout ancient and medieval history the name "fossil" was applied to any mysterious and distinctive object dug up from the earth. Fossils were mysterious in many ways. Although a great number of them bear obvious resemblances to parts of living things — to seashells, leaves, pieces of wood, bones of animals, and even shark's teeth — they are composed of stone instead of organic substance, and some fossils differ in detail from the living things they resemble. If they are really the remains of organisms, then it must be assumed that life forms have changed throughout time. But this conclusion was hard for the learned men of those days to accept because it contradicted the biblical teaching that the earth and all the creatures that inhabit it were created in six days at the beginning of the world. Change seemed to imply imperfection and incompleteness in the design of the original creation.

The position of fossils was another enigma. How could seashells have been deposited on mountaintops? In fact they are not just scattered loosely in the soil. They are often embedded deeply in solid rock as though they have grown there.

The natural philosophers of the Renaissance proposed a number of theories that were hotly debated for several centuries. Fossils are not remains of living things at all, declared one school of philosophers. The fossils have grown within the earth from a characteristic "seed" that contained the potential of a specific form. If a seed of a fish, for example, were deposited in the ground by percolating groundwater, a fishlike fossil might grow from stony material and it would resemble in shape the fish that grew from living tissue.

This theory explained the mysteries of position and composition in the simplest possible way. And in fairness to its advocates, we should remember that the classification of fossil objects as understood at that time included many inorganic substances that actually do take shape within the earth: lodestone, gypsum, mica, and crystalline formations that closely resemble fossil impressions of living plants.

But some early scientists were not convinced by this explanation. Leonardo da Vinci (1452–1519) set down in his unpublished notebooks several reasons for believing that fossil shells are organic in origin. He recognized that the similarities between the fossils and living organisms are so detailed and so precise that a causal relationship is almost inescapable. Leonardo observed that fossil shells are identical with their living counterparts in many special ways. For example, they are preserved in various stages of growth. In some cases other organisms are attached to them or have bored into them. These features reproduce in a striking manner the life situations that Leonardo had observed. The fact that these fossils are often found embedded in rock could be very simply explained, Leonardo said. He had seen how layers of earth are laid down by silting and sedimentation. He postulated that these layers became consolidated into rock by a process of "drying out."

However, Leonardo did not offer any satisfactory explanation of how the shells were deposited in locations now high up in the mountains. He rejected a theory in great favor at that time: that the waters of the biblical deluge had transported the shells up to these great heights and left them there when it receded. Most Renaissance philosophers disagreed; for several centuries after Leonardo's time the deluge theory was generally accepted as the most plausible explanation of the strange distribution of fossils around the earth.

In 1672 an English scientist, Robert Hooke, published a book of drawings based on his observations of fossils and many other natural objects under the newly invented microscope. Like Leonardo, Hooke thought that the evidence for an organic origin of fossils was overwhelming. He also noted, however, that some fossils did not exactly resemble any known living thing and he argued persuasively that some species must have become extinct. The presence of ancient species now extinct could be used, he suggested, to establish the age of the strata in which they appear. In this way the history of the earth could be reconstructed just as the discovery of ancient coins is used to reconstruct the history of mankind. This brilliant thought fell on barren ground. The use of fossils as a dating tool was not developed until almost one hundred and fifty years later.

Hooke also set forth a complete theory of the evolution of the earth throughout time. Beginning as a hot, fluid ball, it has gradually undergone a process of cooling, crumbling, and decay. Earthquakes had been very frequent and severe in early history, he said, although they still continue to disrupt the surface features. The great upheavals and changes in sea level, Hooke suggested, account for the fact that remains of sea organisms are found on mountaintops. Thus the whole history of the earth was portrayed as a gradual deterioration from a pristine and perfect beginning to the world in its present state.

Early in the nineteenth century the distinguished French anatomist Georges Cuvier studied the fossils in several sedimentary layers near Paris and found striking evidence that some species had become extinct. Whole communities of organisms had disappeared, and these events seemed to have occurred rather suddenly because a stratum containing one combination of species was overlaid by another stratum containing an entirely different community. The strata had been laid down in alternating fresh-water and marine conditions. In order to explain these facts Cuvier theorized that a series of "catastrophes" had punctuated the history of the earth, interrupting long periods of tranquil, uneventful history. The catastrophes were correlated with episodes of mountain building, great changes in sea level, and general crustal upheaval. All the disruptions and changes, however, Cuvier declared, were only superficial and cyclic in character. At the end of each period the world returned to its original condition.

This interpretation provided a convenient resolution of the conflict between the religious belief in a perfect original creation and the scientific evidence that the world has been very different at various times in the past. Changes occur, according to this theory, in long rhythmic variations as day follows night, as tides rise and fall, as spring flowers follow winter snows. These fluctuations are transitory — they disappear when seen in the long view of history where a thousand years are "but as yesterday when it is past and as a watch in the night."

When Cuvier's book was published in England, the young geologist Charles Lyell read it with great interest. He liked the theory of cyclic changes, but he argued that it was unnecessary to postulate catastrophes to account for the apparent abrupt changes in the characteristics of adjacent strata. Given a long enough span of geologic time, whole layers of intervening sediment could have been removed by erosion, Lyell pointed out. In fact, everything in geology could be explained on the basis of processes that we see operating in the world today. Carrying the argument one step further, Lyell postulated that these processes have always operated at the same *rates* as they do today. There has been no net change in the earth or in life throughout the portion of earth history that is accessible to scientific investigation. The philosophy based on these principles became known as "uniformitarianism." It dominated geologic thinking throughout the next century and is still influential today.

❧

During all these years a vast store of information had been accumulating: information about the various layers of soil and rock and sand, how they are laid down, how they are uplifted, and how they are finally stripped away by erosion. Many of these discoveries provided support for Lyell's theory that the history of the earth can be seen as one long series of cyclic movements.

Two opposing processes are constantly at work shaping the profile of the planet. One process is driven by forces within the earth that move great blocks of earth crust, cause volcanoes to erupt, make the land shake and buckle, pile up masses of earth material into mountain ranges. The other kind of action is wrought by the force of gravity aided by the wind, the sun, and the rain. Constantly, day after day, it smooths out the landforms, erodes the hills, rounds the sharp peaks. If these forces of denudation were unopposed, they would level the earth back again to something approaching Hooke's original smooth, featureless globe.

Whole mountain ranges as large as the Cascades and the Sierra Nevadas can be created in less than five million years. The breaking down of these landforms after all uplift ceases may take longer — perhaps sixty million years. But that is still a relatively brief period, a short stanza in the long epic poem of geologic time. Actually, most mountain systems are much older than sixty million years because uplift continues even while erosion is taking place.

Rain beating on the slopes washes pebbles and small rock fragments down the slopes. Streams carry them farther down the mountainside, cutting deep gullies as they gather momentum. Winds sweep away the fine particles, depositing them miles away on flatter land. These actions carve a complex maze of steep-walled gorges and narrow ridges.

Even without the influence of wind or flowing water the land slowly disintegrates and moves downhill. Freezing, thawing, and chemical changes cause the rocks to crack and crumble. Released from the stable structure of the parent rock, small pieces roll and slide down the mountainside. In fact, the soil creeps downhill without any visible rolling or sliding. Observation of hillsides over a long period of time shows that the earth moves almost imperceptibly but quite steadily downslope. Earthflows also occur when soil becomes saturated with water. Gravity overcomes the forces of friction and the earth slides down a slope, coming to rest in the nearest gully. As the profile of the land is lowered and softened, the banks and the streambeds become less precipitous. The rate of denudation slows down; a landscape of rolling hills results. But still, the processes continue quietly month after month, year after year. And finally the profile is a gently undulating plain reduced almost to sea level.

Similar downhill movements can be witnessed in a very dramatic way in the icebound world of the high peaks. In this frozen land the rivers become glaciers. The speed of the rushing torrent is turned into a slow but inexorable creep downhill, varying from an inch to nine or ten feet a day. Where the glaciers encounter sudden dropoffs and are confined between precipitous banks, they break up into great jagged blocks of ice like a frozen waterfall. Although the huge chunks of ice may appear quite solid and stationary, an ominous rumbling sound warns the mountaineer that these ice blocks are constantly on the move, grinding against each other. At any moment a more violent upheaval may take place as a large block cracks and shifts its thousand-pound weight to a lower position.

The American Mount Everest Expedition encountered such a chaotic river of ice — the Khumbu Icefall — just above the place where they set up their base camp. Plunging some 2,000 feet between steep mountain walls, this torrent of shattered ice blocks extended for almost a mile and a half across the route to the summit. The leader of the expedition, Norman Dyhrenfurth, described their first attempt to cross this cataract of ice:

> On March 23d two ropes of men moved up through the icefall: Dick Pownall, Ang Pema, and Jake Breitenbach on one, Gil Roberts and Ila Tsering on the other. Suddenly, with a great rumbling, a huge wall of ice toppled over on the men. Gil and Ila escaped injury, but the others disappeared under tons of white debris.
>
> Ten minutes of frantic chopping freed Dick, who was turning blue from a half-ton block of ice pinning his chest so tightly he could hardly breathe. A few minutes later Ang Pema was freed, upside down, badly lacerated, and with a skull fracture. But Jake had been killed instantly and was buried so far back under the enormous wall that no effort could reach him. Some of the team went back up that night and continued searching for his body, but in vain.

In the days and weeks following this tragedy the Sherpa porters of the expedition passed regularly back and forth across the icefall, carrying supplies to camps farther up the mountainside. The hardy people who live in the high mountain villages of the world seem to take on the character of the rugged peaks that overshadow their homes. They develop a stoic acceptance of danger, an understanding of the fragility of human life confronted by the devastating forces of nature. These characteristics are shared by the Sherpas of Nepal and the Quechua Indians of the Peruvian highlands on the other side of the world.

The valleys of the high Andes are among the loveliest places on earth — sheltered oases of green which the equatorial sun fills with golden light as wine fills a cup. In this country of great contrasts the protected mountain regions offer the most favorable living conditions as well as a spectacular setting. The entire coastal area is a barren, sandy desert, bleaker in many places than the Sahara. The land east of the mountains is a lush jungle, turgid with living things, many of which are dangerous to man. Piranhas and alligators and anacondas inhabit the muddy rivers; malarial mosquitoes buzz in the thick humid air; pit vipers and fer-de-lance slide silently along the palm branches and the jungle floor. The soil, leached by heavy rains, is too impoverished to bear good crops. It is not surprising, therefore, that the early inhabitants of Peru settled in the mountain valleys where the air is fresh and where streams from melting snowfields provide a steady source of water. By laborious terracing and irrigation the land was made suitable for agriculture, and green carpets cover the broad staircases that ascend the hillsides. Cuzco, the capital of the Inca Empire, was built at the head of such a valley.

At the northern end of the country the Callejón de Huailas, the "corridor of greenery," lies in the shadow of Nevado Huascarán, the tallest mountain in Peru. This lush valley is so warm and protected that tall royal palms and tropical flowers flourish there at an altitude of 10,000 feet. Rising steeply from the valley floor, the splendid snow-capped bulk of Huascarán fills almost half the sky as if the firmament had opened up and revealed a shining glimpse of heaven. The valley is dotted with towns and villages — Yungay, Ranrahirca, Huarascucho, Calla — spaced out only a mile or so apart along the valley of the Santa River.

This place appears serene and beautiful today. The scars of the dreadful avalanche that happened here are almost invisible now. But nothing can heal the scars left on the hearts of the people who survived the evening of January 10, 1962.

It had been an unseasonably sunny day for early January. The clouds that frequently shroud the peaks at this time of year had not built up to their usual size that afternoon and the sun beat down strongly on the snow-covered range. Glacier 511, one of the many ice fields that glitter along the Cordillera Blanca, lay in a shallow depression just below the very summit of Huascarán. As ice and snow melted, water seeped down through the cracks in the glacier and collected at its base. The moisture provided a slippery surface, reducing the friction of the ice against the rocky face. Gravity — long held in check by friction — was suddenly victorious. The huge mass of ice broke loose from its precarious perch. It plummeted silently in free fall for a brief instant. One of the villagers happened to look up at that moment and saw what looked like a falling cloud. Then "with a roar like that of ten thousand wild beasts" the glacier crashed into a deep gorge. Breaking loose again, it ricocheted down the sheer mountainside, bouncing off the precipitous rock walls. As it moved it gathered momentum, sweeping everything before it: sheep and goats, the roofs of houses, trees, and huge boulders. In less than seven minutes from the time it had broken free, the roaring avalanche reached the Callejón de Huailas, following the course of the Santa River. Three million tons of material, traveling a mile a minute, bore down on the town of Ranrahirca, which lay directly in its path. And in just a few seconds it snuffed out 3,500 lives.

Some of the victims were knocked down by the fierce winds preceding the avalanche and were then engulfed in the wall of mud at its leading face. Many more were stoned to death or crushed in their fragile houses by great blocks of ice and granite. Incredibly mangled bodies were dug out later from beneath as much as sixty feet of debris. The horror struck with unbelievable speed and the finality of total destruction. In several minutes the roar of the avalanche had passed on down the mountainside and was gone. A fine yellow dust settled on the vast wasteland of mud and ice. Those few fortunate people who happened to be standing on high ground away from the river were so shocked they could not speak or cry out or utter a prayer. A profound silence reigned in the valley of the shadow of death.

～

Avalanche and icefall, soilcreep and earthflow — these are just a few of the many ways in which natural forces are gradually ironing out the landforms of the planet. Stone by stone they are tearing down the great mountain ranges — the towering peaks of the Andes, the great granite domes and cliffs of the Sierra Nevada, the frozen summits of the Himalayas — depositing their substance in river deltas and continental shelves and on the floor of the deep sea itself.

It is estimated that North America is now being denuded at a rate that could level a landmass of this size in a mere ten million years. Assuming that this average rate is valid worldwide, such a wiping out of the profile of the planet could have happened ten times since the Mid-Cretaceous Period — just 100 million years ago, when the white cliffs of Dover were formed and dinosaurs left their heavy footprints on every continent (see page 267, Table of Geologic Time). If all the land above the sea were eroded to present sea level, and the remains spread evenly over the ocean floor, it would create a layer about 1,000 feet thick. In 100 million years at this same rate the sediment would be almost two miles deep.

A consideration of these facts raises a number of interesting questions. If so much soil is moved continuously from the land into the sea, we might expect that the ocean basins would be all filled in and the water spread out over the low plains. But the ocean bottoms, on the average, lie a mile or so below the flat land of the continents. Measurement of the deposits on the ocean floor has revealed that they average about 2,000 feet thick, an amount equal to the erosion of the present-sized continents only twice. Although the seafloors are very young compared with the continents (averaging about 100 million years old), we should be able to find five times as much continental residue on the ocean bottom as we do find there. To account for this discrepancy we must assume that the rate of erosion was much less in the past or that the volume of land above sea level has increased very significantly since Mid-Cretaceous times.

Both of these explanations imply real long-term changes in the character of the earth. It has not returned to its original state after each cycle of uplift and erosion. For nearly four billion years the competing forces have tugged at the surface of the planet, carving ever-changing landscapes where shining mountain summits are gilded in the dawn's first light, and glacial torrents carry the soil down again through rainbow canyons to the sea.

This ceaseless to and fro does not repeat the past. It is only a superficial rhythm impressed upon some long-sustained process of becoming. Like a child lulled to sleep, the planet evolves, grows older, discovers new potentials, even while it is rocked in the cradle of time.

The Hidden Land

There is, one knows not what sweet
mystery about this sea, whose gently
awful stirrings seem to speak of
some hidden soul beneath.

— HERMAN MELVILLE,
Moby Dick

Human beings have always known instinctively that something awesome — something magnificent and at the same time frightening and humbling — is held in the sea's embrace. The ocean guards well the mysteries of this underwater world. The salty waters that lap the shores of every land are well designed for keeping secrets. Look out across the sea on a bright sunny day, observe how much the ocean surface tells — and how little. Like a half-silvered mirror the sea reflects the moving image of the great cloud masses that float high in the troposphere carried by the winds aloft. Its surface glints and laughs with vivacious patterns of gold and blue. These little waves sing the tune of vagrant breezes that ripple across the surface, while the long slow rollers beat out the rhythm of storms that rage beyond the horizon. The currents reveal the turning of the earth in space and the changing of the seasons. The tides rise and fall like the slow, regular breathing of a sleeping giant as the moon circles around the earth and draws the water across the ocean basins.

These signs all yield vast stores of information about the winds and the weather, the motion of the planet and its satellite around the sun.

But the face of the deep gives only the slightest hints about the hidden landscape and the creatures that dwell in its depths: a shadow that does not move, a patch of paler blue, a line of breakers far out at sea. These are only faint echoes of that rich and incredibly diverse land below the shining surface.

Our remote vertebrate ancestors inhabited this most majestic landscape on the planet's surface. Like a vague childhood memory, felt but not understood, they carried the imprint of it with them when they left their birthplace and set forth to conquer a new world. Through all the eons that have passed, this imprint — like the salt water that runs in our veins — has not been lost. Three hundred and fifty million years — perhaps two billion generations — later, mankind carries within him this love and awe of the sea.

In early human societies it found expression in religions, legends, and fairy tales. Imagination peopled the deep with angry gods, dragons and monsters so voracious that they would devour ships whole, seductive sirens who lured sailors to their death. The truth as it has been revealed is much more wonderful than any of these fairy tales invented before we could explore the bottom of the sea, before we could float beside the soaring mountain walls, move along the deep canyons, and watch new crust boiling up from the hot interior of the planet.

Before we could explore the bottom of the sea — that little phrase encompasses most of human history. Until early in this century two thirds of the earth's surface was almost completely unknown, for the ocean did not give up its secrets easily to those who had left its sheltering care, had taken on new ways, and had lost the art of survival in water.

~❧

The first men systematically to penetrate the dimly lighted depths were sponge divers. As early as Homer's time in the ninth century B.C. sponges were sold and traded commercially. They were used for cleaning purposes just as they are today, and also for more exotic needs. Soldiers used them to pad their heavy helmets; robbers attached them to the soles of their sandals in order to move noiselessly. In the Aegean sponges were plentiful, growing on the bottom of the sea at depths of between seventy-five and a hundred feet. Sponges were gathered there without any artificial breathing apparatus or protection against the water pressure that is experienced at these depths.

The divers were taken by boat to a place where the sponges were known to be abundant. As the anchor was lowered, the divers stared down into the lowering face of the sea, looking anxiously for some good omen that this day, at least, they would come to no harm. The sight of a certain kind of fish called "anthias" by early writers but now believed to be a grouper was taken as a good augury. "In waters frequented by the anthias," said Aristotle, "it is quite certain that there will be no dangerous fish. The sponge divers therefore regard its presence as an indication that they may safely enter the water, and they hold the anthias in veneration as a sacred fish."

Even without a favorable omen, however, the sponge diver had to prepare for his descent. He attached a rope around his waist, poured oil in his mouth and ear canals. Then he soaked two sponges with oil and placed these over his ears. Taking in his hands a very sharp curved knife and a heavy stone, he inhaled deeply and dove. The weight of the stone carried him down rapidly to depths where golden sunlight faded to a soft, murky blue twilight and the pressure on every square inch of his body was three times as great as at sea level. The diver's ears, even protected by the oil sponges, began to cause great pain. His head, eyes, and chest felt as though they were being pressed in a powerful vise. When he reached the bottom, he spit out the oil from his mouth. It rose to the surface, calming the waves, and allowing more light to penetrate into the depths. Then, working as quickly as he could, the diver cut a sponge or two loose from the rocks on which they were growing and, tugging on the cord attached to his waist, he gave the signal for his companions in the boat to pull him back to the surface. As he rose through the water, the sponges bled a nauseating liquid that spread around the diver. Eager to end this ordeal, he ascended as rapidly as possible, thus giving his body no chance to adapt to the rapidly decreasing pressure. After he was pulled back into the boat he shook water and often blood, too, from his ears and nostrils. Broken eardrums were common and it was reported that the divers slit their nostrils in order to breathe more freely.

"No ordeal is more terrible than that of the sponge diver," said the Greek poet Oppian. "Sometimes his efforts end miserably and cruelly. Once he has plunged into the waves the unfortunate often does not return. He has encountered some hidden monster of the deep and lost his life. Desperately at first he pulls the rope in order to be hoisted to safety, but the monster seizes him and then a horrible tug of war takes place. The monster holds him from below and his friends pull him from above, disputing the half-devoured corpse of the unfortunate diver between them."

Such terrible dangers accompanied man's first serious attempt to enter the underwater world. The sponge diver had no time or inclination to look around him and observe the beauty of this hidden land. Driven by necessity and custom, he continued day after day to brave the same perils. The divers frequently made forty to fifty plunges in one day. Diving for sponges and also for pearls was the best, perhaps the only, way of making a living in many communities. In several countries — notably India, Ceylon, and Japan — women also became expert divers. They were usually able to stay under water longer than men. In fact, pearl diving became traditionally a woman's occupation.

The use of a breathing apparatus to aid the underwater swimmer was first exploited, as many scientific inventions have been, in time of war. In 480 B.C. Xerxes launched the second major Persian attack on Greece with what has been described as the largest army ever to have been assembled before our own century. Herodotus reported that there were 2,641,000 fighting men and an equal number of engineers, provisioners, slaves, and prostitutes. Wherever the army ate two meals, the city that fed it was utterly ruined. This vast horde, supported by a fleet of 1,207 transports and fighting ships, crossed the Hellespont on a temporary bridge of ships and swept south along the Greek peninsula toward Athens and Sparta. As word of their approach spread before them, terror and desperation gripped the little Greek states. Effective resistance seemed impossible, but surrender was unthinkable. The Athenians hastily gathered together their tiny fleet, numbering only a fraction of the Persian warships, and sailed it north to challenge Xerxes' approaching force.

In a bay along the coast of Thessaly near Mount Pelion, a man named Scyllias and his young daughter, Cyana, saw a great number of Persian ships riding at anchor one spring day. The boats were straining at their moorings, buffeted by winds as a storm approached from the west. Scyllias and Cyana recognized a great opportunity to strike a decisive blow against the invading force. They plunged into the sea, holding hollow reeds in their mouths, and swam underwater so just the tips of the reeds broke the surface, making tiny V-shaped waves that quickly disappeared. Fighting their way slowly through the increasingly turbulent water, the two swimmers finally arrived unobserved in the center of the great Persian fleet. As thunderheads scudded across the sky and the rising wind whined in the rigging of the sailing ships, the two silent swimmers made their way from boat to boat, casting off the anchor lines and setting the boats adrift at the mercy of the storm. Many warships were lost in the confusion that ensued.

A few days later the Athenian fleet engaged a somewhat reduced Persian navy at Artesium and Salamis and won important victories. Historians agree that the whole course of Western civilization would have been altered if the Persians had succeeded in conquering Greece. At this great remove and in the face of conflicting details in the reports of the exploit, it is impossible to estimate just how large a debt we owe to the underwater warriors Scyllias and Cyana. But the Greeks themselves were so thankful that statues were commissioned in honor of the swimmers and were set up in a temple at Delphi.

In the meantime the remnant of the Persian fleet fled to the Hellespont where Xerxes, in the wake of another storm, is said to have punished the sea with three hundred strokes of the lash laid upon the waves. It was also reported that he sent men to brand it with red-hot irons, and he shouted these arrogant words: "Thou bitter water, thy lord lays on thee this punishment because thou hast wronged him without a cause, having suffered no evil at his hands."

❧

Other wartime uses of the snorkel tube are mentioned by ancient writers. Spies penetrated enemy lines by swimming underwater at night; swimmers carried provisions in skin bags to relieve besieged cities and drilled holes in the hulls of enemy warships riding at anchor. But snorkel tubes allow access only to water just below the surface. Exploration of the deeper world required some method of taking a supply of air down with the diver.

One of the first inventions of this kind was described in the works of Aristotle. A cauldron was placed over the diver's body and forced straight down into deep water, thus trapping a bubble of air under pressure at depth. Many versions of this idea were developed throughout the ensuing centuries, varying from small glass globes to iron bells, airtight barrels, and huge caissons. These devices were useful for salvage and underwater construction, but they had their limitations, too. They did not allow the diver to move around freely and explore the ocean floor.

Leonardo da Vinci, who had original ideas about almost everything, drew up plans for a diving helmet studded with sharp spikes to ward off dangerous sea creatures. It was also equipped with glass-covered eyeholes, and a breathing tube that reached to the surface, like a long snorkel. This diving apparatus was never built — a fact that may have been fortunate because the pressure on a person's lungs just a few feet below the surface makes it impossible to draw air in through a long tube of this kind. However, the webbed gloves and flippers sketched by Leonardo would have been a valuable addition to diving equipment and were not reinvented for many centuries.

Extensive exploration of the underwater world had to wait until certain key problems had been solved. The effect of pressure on the human body was not understood until the late nineteenth century when a French physician, Paul Bert, proved that nitrogen, which comprises 78 percent of the air we breathe, dissolves in the blood when the body is under great pressure and forms tiny bubbles if the pressure is relieved rapidly. These bubbles cause "the bends" — an agonizing condition that sometimes results in paralysis and death. Bert demonstrated that slow decompression allows the nitrogen to escape without forming bubbles.

Ways of protecting the diver against cold were also necessary. Even in warm surface water the body loses heat much more rapidly than it does in air of that same temperature. Modern wet and dry suits provide some protection against heat loss. Really deep ocean water throughout the world, including the tropics, hovers just a few degrees above freezing. At these depths enclosed vessels with windows for observing the underwater world offer a completely warmed and pressurized environment. Gradually the technology was worked out that led to the diving suits and submersible craft and decompression tanks that have made possible the exploration of this last great frontier.

~✔

In addition to the techniques for taking man himself down into the sea, many instruments have been invented for obtaining information about the ocean floor from ships on the surface. Several of them were developed and perfected for use during wartime. The precision depth recorder is an echo-sounding system invented for detection of enemy submarines, submerged reefs, and other hazards concealed beneath the sea. Pulses of sound are directed down to the ocean floor and reflected off the terrain back to the ship, where they are picked up by an automatic device that calculates and records the time taken for the round trip. This information is plotted continuously, drawing a profile of the ocean floor. These instruments are now standard equipment for oceanographic research and are run continuously as the vessel cruises across the ocean.

Commerce, too, has made important contributions, providing financing for the development of sophisticated scientific instruments. The oil-drilling industry pioneered the technology that led to the development of deep-sea drilling, one of the oceanographer's most remarkable tools. Sampling of the ocean bottom — a procedure that for many centuries was performed with dredges and cables — is now a finely perfected art. An oil-drilling rig is mounted onboard a specially designed ship, the *Glomar Challenger*, which is equipped with computer-controlled stabilizers and water jets that maintain the ship's position and orientation with a remarkable degree of accuracy even when sampling is conducted in heavy seas. The long, flexible drill can collect core samples in water up to four miles deep and it can bore through several thousand feet of crust. The rocks and sediment brought up by the drill are studied in many ways. The layering of the core gives significant information about the historical development of the seafloor. The rocks are examined under a microscope; their crystal structure and chemistry are analyzed; and their age is measured by radioactive dating.

This technique of measuring age depends upon the fact that certain radioactive elements may be incorporated into rocks when they solidify from the liquid state. As rocks age, these elements disintegrate, each element having its own characteristic rate of decay, and the stable products accumulate, trapped within the crystal structure. By measuring the amount of parent material and daughter product present, and applying the known disintegration rate, the time elapsed since the formation of the rock can be estimated.

When the facts from all these sources as well as the photographs and samples taken by divers and submersible craft are arranged in a composite picture, an astonishing portrait emerges of the land that lies beneath the surface of the sea.

~&

If all the water were drained out of the oceans, a striking landscape would be revealed. A tremendous mountain chain bisects the Atlantic Ocean and continues in an unbroken range running in a broad loop between Antarctica and the Cape of Good Hope, up through the center of the Indian Ocean, then down around Australia and through the South Pacific to Mexico. It is almost 30,000 miles long and it averages 600 miles wide, occupying nearly one fourth of the whole surface of the planet. In addition to the long, continuous mountain range there are many broad stretches of wildly distorted terrain where volcanic cones occur in haphazard arrangement; deep gorges and canyons cut wide swaths across the scenery. The floor of the western Pacific is such a region, occupying more earth surface than the whole Atlantic Ocean. (See the map of the ocean floor in Figure 1.)

If we could go on a walking safari through this land from which all the sea had been drained, we would see some sights reminiscent of continental landscapes and some as alien as a landscape on the moon. Many of the isolated mountain peaks are crowned with white as though they have been dusted with a fine powder snow, and — just as in the Alps or the Himalayas — there is a "snow line," an altitude below which the snow seems to have melted away. This impression is not far removed from the truth. The white powder is a fine sediment composed of the tiny shells of trillions of microscopic organisms. This sea-snow drifts softly down day after day, collecting on the high slopes. In the colder and more acidic waters at greater depth, it dissolves in the seawater.

Other types of sediment, darker in color, blanket the lower slopes and collect in the flat abyssal plains. These are composed of clay and silica sand washed down from the continents and soft ooze derived from the insoluble siliceous remains of aquatic life. Spread out in a thick layer, they have gradually consolidated into a firm flat surface that covers irregularities of terrain, thus creating a great sameness. This featureless landscape is just what men expected to find across most of the ocean floor but actually it occupies only a small fraction of the whole. Most of the land is broken up with canyons and sea mounts and lakes of hardened lava. Occasionally there are hot geysers and multicolored metallic deposits. It is a sight-seeing trip that would make a hike through Yellowstone Park seem dull in comparison.

If we begin our trip near the Mid-Atlantic Rift range, we see it rising abruptly out of the abyssal plains, stretching across the entire horizon — a dramatic mountain wall with snow-covered slopes. Along the very center, on the highest peaks of the range, we observe an interesting effect: no snow has collected there. Dark black rocks stand out prominently along the crest, and then on either side a light dusting begins to appear. Farther from the crest in both directions the sediment becomes deeper. Here and there it has accumulated in little depressions, making small isolated drifts on the mountainsides.

As we climb the range, we reach the point where the sediment becomes very thin and then the shining rock surfaces stand out sharply. Some are fractured and heaved up into steep walls. At close range it is apparent that they are composed of basalt, one of the rocks that characteristically forms from molten lava when it cools rapidly. There are mounds of rounded basaltic rocks with glassy smooth surfaces. "Pillow lavas" like these have been observed to form in places such as Hawaii, where lava erupts beneath the sea and solidifies quickly in cold water, preserving the extruded shape like ribbons of toothpaste squeezed from an enormous tube.

At the top of the range we find a deep valley forming an enormous trough that runs the length of the crest. It is filled with heaps of dark rock and looks very much like the inside of any recently active volcano. But instead of the typical cone shape, the mouth of this volcano is stretched out in one dimension, making a rift valley extending in the same direction as far as the eye can see.

When samples taken from various locations on the range were dated, it was discovered that the rocks along this rift valley are among the youngest pieces of earth crust: those from the center of the rift itself are usually less than a thousand years old and the rocks become progressively more ancient in places farther from the ridge. The young rocks do not occur only at the surface as would be the case on the lava fields of Mount St. Helens. Here on the ocean rift they extend deep into the crust, as core samples have demonstrated.

Now we can understand the absence of "snow" on the high ridge.

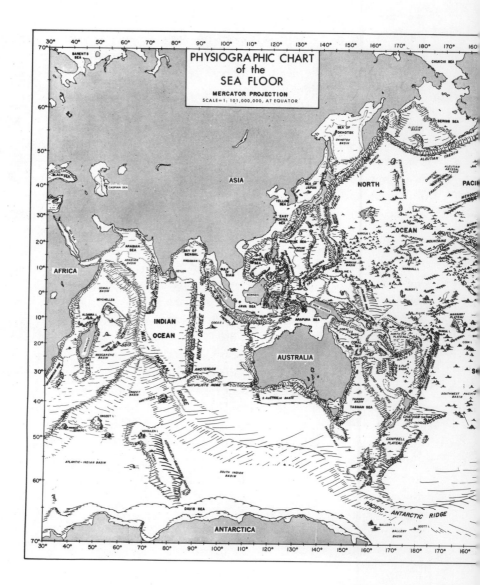

PHYSIOGRAPHIC CHART
of the
SEA FLOOR

MERCATOR PROJECTION
SCALE=1; 101,000,000, AT EQUATOR

34

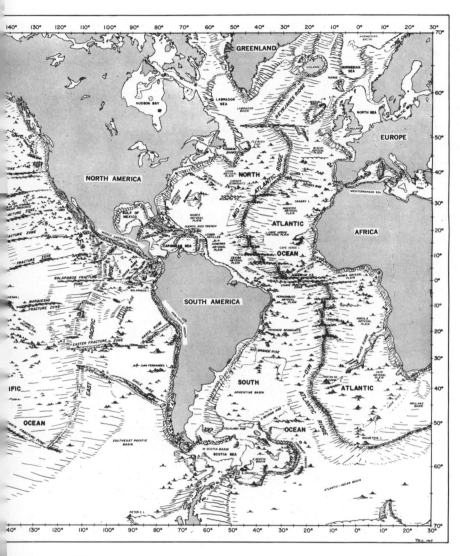

FIGURE I   *The Sea Floor. (Reprinted with permission of Hubbard Scientific Company)*

The rocks here have not existed long enough to collect a layer of sediment. The sea-snow falls slowly and gently, usually accumulating less than a tenth of an inch in a thousand years.

An aerial view of the great range would show the long rift lines curving from horizon to horizon. The edges of the rift are parallel but offset a little to the left or the right in many places, making a wavy pattern when see from far away, like the jaws of a giant clam. Between them a deep fissure leads down to the soft interior, where something is stirring within — "gently awful" stirring, as Melville wrote about the sea long before the rift was known, seeming "to speak of some hidden soul beneath."

Soul? Or the soft inside body of the planet? Analogies fill our minds and are rejected one by one. Something is happening here that is unique in our experience. The earth's crust is spreading apart along these gaping cracks in the planet's surface. Although the rate of spreading is very small in human terms — just a few inches a year — inexorably, eon by eon, as geologic clocks tick, it takes place fast enough to have made anew the entire ocean floor, about two thirds of the planet's surface, twenty times since the earth was born.

Clambering down from the high range and taking care to skirt around the talus piles of loose rocks that have accumulated near the foot of the range, we continue our tour of the hidden land. Traversing the flatter portions of the seabed, we encounter many square miles that are studded with dark stones like little lumps of charcoal or cinders. These are known as manganese nodules, but they are also rich in other important minerals: copper, nickel, cobalt, zinc, molybdenum, lead, vanadium, titanium. The reasons for the formation of these interesting stones are not well understood. Many scientists believe that the process involves the interaction of submarine volcanism and seawater. Metallic oxide molecules may precipitate around some nucleus, some tiny particle of shell or sediment, in much the same way as raindrops precipitate around dust particles in the atmosphere. The nodules accrete in concentric layers, making an onionlike structure. Manganese nodules form in places where water contains abundant oxygen. They are growing today on the beds of lakes in Sweden and have been mined there since the middle of the nineteenth century. These nodules on the seafloor may become one of the most valuable resources of the oceans. An estimated sixteen million tons are forming there every year — a storehouse of rare and important minerals manufactured annually by the ocean.

As we proceed over vast distances of seascape, we encounter many evidences of volcanic activity. We must be constantly on the alert for eruptions of hot lava and gases from fissures beneath our feet. Little volcanic cones twelve to fifteen feet high, looking like termite mounds, are festooned with lava that has poured down from the top and congealed along the sides of the cones. Large volcanic mountains rise abruptly like so many giant stalagmites from the seafloor. We walk across lakes of hardened lava that appear to have solidified rapidly, freezing swirled patterns into the hardened surface.

Deep-sea drilling has revealed the surprising fact that the ocean floor is almost entirely volcanic in origin. The underlying base below any accumulations of sediment is composed of basalt. Furthermore, the ocean crust is relatively thin. It is only four to five miles thick. The crust of the continents, on the other hand, typically varies from twenty to thirty-five miles in thickness.

The entire ocean floor is very young compared with the continental crust. Rocks 3.8 billion years old have been found on land, but no sample of ocean floor older than 180 million years has been discovered. Some of the oldest portions lie on either side of the Atlantic, off the east coast of the United States and the west coast of Morocco. There is also a portion of similar age in the north central Pacific.

In geologic perspective 180 million years ago is the comparatively recent past. At that time mammals had already evolved and had taken up residence on every continent. This was the heyday of the enormous reptiles. *Brontosaurus*, *Stegosaurus*, and *Brachiosaurus* roamed the land masses of the earth. So the crust of the ocean floor is younger than the bones of many dinosaurs that we see displayed in museums around the world.

As we continue our tour of the hidden land we come quite suddenly upon a very deep valley in the earth's crust. The land falls away to a depth of several miles, creating a more dramatic landscape than the view from the top of Mount Everest. There are many such fissures in the earth's crust. The deepest ones occur in the Pacific, near the

Marianas, the Philippines, the Aleutians. Theories about them form an important part of our story, and they will be considered in some detail in Chapter 4.

When we approach a continent after walking across many miles of abyssal plains, circumnavigating volcanoes and avoiding the deepest trenches, we come finally to a place where a very gentle rise of land begins and continues quite uniformly for several days' march. Then all at once we see ahead of us a long high bank of earth rising about a mile in altitude. The gradient of this enbankment is relatively gradual, but it rises to greater heights than any escarpment on the continents and its face is deeply scarred with enormous erosion channels. They look like canyons cut by rushing rivers and, indeed, they are believed to have formed in that way. The top of the slope is the real edge of the continent. At certain periods in the earth's history when glaciers locked up large amounts of the planet's water supply, ocean levels were lower and this flat plain, the continental shelf, was dry land. Rivers cut valleys on the brink of the shelf as they disgorged their water into the sea. Then when the glaciers melted and the ocean levels rose, the continental shelves were gradually inundated. But underwater currents continued to use the old channels, cutting them deeper and deeper as they brought their dense mixtures of water and sediment from the continents down to the deep seafloor.

Climbing the slope onto the shelf is a dangerous and difficult undertaking. The sides of the canyons are precipitous and the soil gives way easily to slumping. One false step may start a great mud avalanche. But when we finally reach the top, we can enjoy a wide view of gently rolling land. It is intersected by many gullies and molded by bottom currents into hillocks that resemble sand dunes on the desert. The repetitive rise and fall of sea level has left traces of old beaches, deposits of gravel and mud. Fossilized bones of large land mammals — mastodons, mammoths, and giant sloths — have been uncovered here, as well as beds of peat containing fossils of grasses, twigs, and pollen. This testimony from the past confirms the assumption that the continental shelves are really drowned portions of the continents. Further evidence has been discovered by the Deep Sea Drilling Project. Cores taken through the sediment and rock show that the shelves lie on a granitic base like the continents, not on the basaltic crust that is characteristic of the ocean floor. Over the basement rocks lie great accumulations of soil washed down from higher land.

All the continents on earth are surrounded by shelves that vary greatly in width and depth. The depth may be anywhere from 30 to 2,000 feet. The width varies from less than half a mile off the west coast of Peru to more than 800 miles off Siberia. Taken all together, the continental shelves occupy one tenth of the world's surface. Commercially they are very important because they harbor 90 percent of the earth's fish resources and important stores of oil and gas.

We have been looking at the bare bones of the hidden land, stripped of its soft fluid medium, emptied of its vast panorama of movement, color, and life. In order to grasp the full impact of the ocean sea, we must fill the basins back up with 48 billion billion cubic feet of salt water and 250,000 species of living things. We must put the schools of flying fish back in the sapphire waters of the Caribbean, restore the humpback whales to their breeding grounds south of the Bering Sea, plant the flowerlike sea anemones back on the coral reefs of the Pacific, decorate the shoals of the Indian Ocean again with butterfly and angel fish. Once more the great body of the sea leans heavily upon the hidden lands out to the farthest margins.

Water has several special characteristics that help to make our environment benign. It absorbs and loses heat very slowly, tempering the heat in summer and the cold in winter. Most substances contract when cooled. Water does this, too, but not quite to the freezing point. Just before it reaches that critical temperature it begins to expand and when freezing occurs the expansion is dramatic. For this reason, ice is less dense than water. It floats, remaining on the surface and forming a protective, insulating shield that slows down further freezing. This unique behavior moderates our climate and prevents large bodies of water from freezing solid; so even in the Arctic the seas can sustain a rich and varied community of aquatic life. Water has the capacity to dissolve and carry in solution more substances than does any other liquid. A great diversity of compounds and elements are circulated by the currents, essential nutrients are made available and widely distributed around the earth.

In small quantities water looks transparent but actually it turns back much of the light that falls upon it and hungrily swallows up the portion that enters its surface. Blue rays are transmitted best through water, but they are scattered, giving the sea its beautiful aquamarine color. In shallow waters such as those on the Bahama Banks the bottom sands are still drenched with light. The water ripples over them, transparent as pale Venetian glass. But as the depth increases, less and less of the light is able to penetrate. The red and orange rays of sunlight are totally absorbed by a hundred feet of water and most other visible light by four hundred feet. At this point the underwater world is immersed in a sapphire blue twilight. Deeper still, that light, too, fades and finally at a depth of 1,000 feet no sunlight penetrates. There is velvety darkness, an everlasting nighttime that never varies from hour to hour or season to season. The only illumination that relieves the gloom is made by the little sea creatures themselves, giving off a soft phosphorescent glow, like the glimmer of fireflies on a distant hill.

The ocean water at depths of about one or two miles is extremely cold — just a degree or two above freezing — and pressures are hundreds of times those experienced at sea level. For many centuries it was believed that this forbidding environment could not harbor any living things. But when submersibles descended to these regions, they discovered a varied community of aquatic life. They found deep-sea sponges and crabs, sea spiders and sea cucumbers. Bizarre fish also inhabit these depths: the ugly viper fish with sharp curved fangs, the hatchet fish with a long whiplike tail. Weird apparitions like these have been viewed and photographed through the windows of submersible craft.

During the years 1973–1976 a cooperative scientific venture called Project FAMOUS (French-American Mid-Ocean Undersea Study) explored a portion of the Mid-Atlantic Rift, using both surface and underwater craft. There were three submersibles: a large French bathyscope named *Archimède*, built to withstand pressures at the ocean's greatest depths; a more delicate little French diving saucer named *Cyana*; and an American submersible, *Alvin*, which is intermediate in size and capability between the two French vessels. *Alvin* can descend with a comfortable margin of safety to 12,000 feet. It has long mechanical arms to pick up rock samples from the ocean floor. Two stereo cameras mounted on the outside of the vessel are able to take pictures every ten seconds, providing a continuous panorama of the ocean landscape.

The scientists who rode in *Alvin* were deeply impressed by the wild loveliness of the scenes in the rift valley. "The seafloor is so beautiful, so out-of-this-world beautiful," said Tjeerd Van Andel. "The misty, grandiose, mysterious landscape of craggy black and snowy, pillowy white . . . is unforgettable."

Evidence of violent fracturing and cracking was apparent almost everywhere the scientists looked. Along the central axis fresh lava flows were found interrupted occasionally by very fine hairline cracks. With increasing distance from the axis these long fissures widened in some places to twenty or thirty feet, creating vertical fault scarps. The American geologists were anxious to explore these fissures, hoping to find evidence of hot springs and deposits of minerals such as those reported by *Archimède* and *Cyana* slightly farther north.

~⌘

Dive 526 of *Alvin*'s career began in a routine manner. Many samples had been collected in the hopper and it was almost time to return to the mother ship, *Lulu*. But at that moment they were passing by one of the larger fissures and they decided to take a look inside. Slowly and carefully they edged their craft along the deepened crack. As the fissure descended it also narrowed slightly, and suddenly the pilot, Jack Donnelly, found that he could not move the craft farther forward without damaging the skin of the vessel against the jagged walls of hardened lava. He decided to abandon the exploration of the fissure and tried to start the sub back up, using its lift props. But the vessel was wedged tightly and would not rise. Worst of all, it would not move backward either.

Aboard *Lulu* a mile and a half above *Alvin*, the man who was monitoring the computer plot of *Alvin*'s position saw all movement suddenly come to a halt. Minutes dragged on into half an hour . . . an hour . . . still no sign of motion.

"Better get under way," he prompted over the intercom. "Mission time is running out."

"We're trying," was the alarming response from Jack Donnelly. "We don't seem to be able to rise."

In the control room all light conversation and laughter ceased abruptly as the men on the surface talked with the three men pinned in the narrow crack in the earth 9,000 feet below the surface of the sea. Following a suggestion from the men on top, Donnelly tried wobbling the vessel by rotating the twin lift props. *Alvin* moved slightly — a few inches. Donnelly tried the same technique again and gained a few inches more. After ninety minutes of painstaking maneuvering the craft finally came free, backing slowly out of the narrow fissure. The sun was very low in the sky when *Alvin* rose back up to the surface and the safety of the mother ship.

❧

Since that day *Alvin* has gone on many other expeditions to study dramatic features of the deep seafloor. In the Cayman Trough south of Cuba it carried geologists down into another spreading rift — this one at a depth of 12,000 feet. Along the base of the rift valley's west wall the geologists found rocks that seemed softer than the samples they had retrieved before. When these rocks were examined with hand lenses onboard *Lulu*, the scientists found that they held in their hands peridotite and dunite: rocks composed of minerals that are believed to be the principal constituents of the earth's mantle, the layer that lies below the earth's crust.

Another research team took *Alvin* to the equatorial waters of the Pacific to explore the Galapagos Rift west of Ecuador. On this trip they were searching again for hot springs. Evidence of unusual underwater features in this area had been reported earlier by surface vessels. Temperature variations in the water just above the seafloor suggested the presence of hydrothermal vents in the floor. So the Galapagos Hydrothermal Expedition was undertaken in the winter of 1977.

From a research vessel, cameras and thermometers were lowered toward the floor of the rift valley. The temperatures recorded at this depth were near freezing, but suddenly the temperature chart showed a sharp spike and when the photographs that had been exposed at that same moment were developed they showed a dense blanket of giant clams covering the seafloor. *Alvin* was taken down to examine this re-markable phenomenon at close range. Through the portholes the diving crew looked out on a bizarre oasis of life and warmth set in the cold darkness of the abyss. *Alvin*'s lights revealed fountains of milky blue water pouring from fissures in the ocean crust. The water was warm (about 60° F) and laden with bacteria and tiny particles of sulfur. The rocks bathed by these warm springs were almost solidly encrusted with enormous mussels as large as a man's head. Armies of crabs scurried along the seafloor, feasting on seaworms and other small organisms that grew attached to the rocks. An octopus with fins like large bat ears floated by *Alvin*'s window and a fish swam by, waving a thin hairy tail. A few miles farther along the rift other hot springs were discovered. Each had their own characteristic community of living organisms. Huge tube worms monopolized one vent, which the research team dubbed "The Garden of Eden," while various bivalves were most common at the others.

Samples of the spring water, pieces of rock, and even a few clams were brought to the surface. When they were opened up on deck, a strong smell of rotten eggs filled the air. Hydrogen sulfide — this smelly compound holds the secret of the abundant life in these regions, which are entirely cut off from the energy of sunlight. Under extreme heat and pressure the sulfate normally present in seawater is converted to hydrogen sulfide, a compound in which heat energy is stored. Certain types of bacteria are able to metabolize the sulfide and extract its energy just as animals extract energy by metabolizing sugar, starch, and protein. These bacteria thrive in the warm sulfur springs — a health spa for life in the abyss. They multiply, providing food for larger organisms like the crabs and clams. Thus a food chain is built up, using as its base the heat energy of the earth's mantle instead of the solar energy on which most of the planet's life depends.

The hot springs of the Galapagos, dramatic as they are, pale in comparison with the hydrothermal vents discovered off the coast of Baja California in 1980. At the Galapagos the springs well gently up from crevices in the ridge crest. The maximum temperature is about 62° F, like a mild April day. In the rift off Baja California smoky gray fluid roars upward with tremendous force like jets from a fire hose. The scientists riding in *Alvin* stuck a temperature probe down into the opening of one of these vents and the probe melted. The water temperature as it emerged from the opening was 700° F. These black smokers are laden with a heavy concentration of iron sulfide. At the vents where they emerge from the crust they have built up mineralized chimneys that grow to heights as great as ninety feet before they collapse and shatter on the ocean floor.

The hot springs are created by seawater circulating through the outer layers of the earth like a giant organism breathing in and breathing out. Cold water seeps down through porous rocks until it reaches the heated rock of the mantle. Then, scalding hot, it expands and pushes its way up again with great force. As it rises through the crust, the hot liquid reacts with the rocks and finally spurts forth as a hydrothermal vent laden with minerals and gases. When it strikes the cold seawater again, some of the minerals — iron oxides, sulfides, copper, and manganese — are precipitated on the ocean floor. One of the most surprising discoveries is that many of the vents are emitting bubbles of methane gas and globules of oil. Are midocean rifts acting as natural oil refineries? The new findings present scientists with many new mysteries but at the same time they promise to resolve some of the older ones. For example, geochemists are hopeful that the discovery of hydrothermal vents will help explain the presence of the mysterious fields of manganese nodules that are so widely distributed across the ocean bed.

Another riddle has plagued oceanographers for decades. The chemical composition of seawater is quite different from the water of the rivers that feed it and yet its chemistry is remarkably constant all over the world. If the ocean basins were just passive bowls into which rivers discharged their flow, they would act like immense evaporation pans. Given this simple model, the expected chemical composition of the ocean can be calculated. But the results do not correspond with the known chemistry of seawater. Now with the discovery of hydrothermal activity these differences may be explained. The great fissures with their fountains of hot fluids act like enormous chemical plants, affecting the characteristics of the planet's oceans and atmosphere as well as constructing new material for its crust. Earth, air, and water are continuously altering one another, locked in a closely interdependent relationship that has influenced our world throughout geologic time.

The venting process along the rifts may also throw new light on one of geology's oldest and most tantalizing unanswered questions. Did the same amount of water fill the sea basins, form the clouds in the sky, and bathe the soil in rain when the earth was very young? Almost all scientists agree that water was part of the original components of the infant planet as it condensed out of the cooling solar nebula. Various gases and liquids, adsorbed by the heavier materials, became incorporated into the solid mass as it formed. Under the influence of gravity, heat and pressure built up and water was released from the rocks. But how rapidly did this action take place? Some geologists believe that most of the water was set free early in the planet's history. Others theorize that it has been released more gradually and that the process is still going on to a significant degree today. Hydrothermal vents and volcanoes discharge a great deal of water. Its high temperature proves that it has passed through layers below the earth's crust. A small part of it may be truly new to the planet's surface — "juvenile water." But the rest is recirculated water that has filtered down through the overlying layers of rock, heated on contact with the material of the mantle, and ejected again. What are the true proportions of old and new? More detailed study of the deep-sea jets may help to settle this controversy. The weight of scientific opinion today favors the conclusion that most of the water escaped very early and that the same quantity of water present on the earth's surface today was already here two to three billion years ago.

So the water of the ocean sea may be one of the oldest things on earth, as old as the most ancient rocks of the continents. Like a sophisticated lady getting on a bit in years, the sea does not reveal her true age. We might say that age has no real meaning for her. She is constantly made anew every day — shaped up, rejuvenated, recycled, and refreshed by intimate contact with the land and the winds. No one, not even Xerxes, can lay a mark upon her body that is not quickly erased.

Throughout the planet's history the sea has carved and molded the character of the land. She has scooped out steep escarpments and deep gorges, impressed the rhythm of her movement on the hard rocky shores of the continents. But still she is ever yielding. Beneath her smiling, enigmatic face there are grave depths where silence and darkness dwell always. Here in these hidden places she watches impassively while the earth tears itself violently apart and makes itself anew. Quietly giving way to make room for the growing landmass, she receives and holds this newborn substance in her soft embrace.

3

Lodestone and Star

Magnetic force is animate, or imitates life;
and in many things surpasses human life.

— WILLIAM GILBERT, *De Magnete*

The magnetic field of the earth is the most truly mysterious property of the planet. It cannot be seen or heard or felt (except under very special circumstances). But it turns a magnetized needle and establishes a unique direction at every point on the earth's surface. The imprint it has left in the rocks has provided important clues to the history of the planet. And yet the more we have learned about it, the more it defies explanation. It is, as geologist Allen Cox remarked, "one of the best-described and least understood of all planetary phenomena."

The ability of certain dark, heavy stones to attract iron was discovered in Greece long before the Christian era. It was mentioned by Thales of Miletus (624–548 B.C.) and Plato (427–347 B.C.). The Roman poet Lucretius described it in his *De Rerum Natura*, written in the first century B.C.: "Iron can be drawn by that stone which the Greeks call Magnet by its native name, because it has its origin in the hereditary bounds of the Magnetes [the inhabitants of Magnesia in Thessaly]. . . . Sometimes, too, iron draws back from this stone; for it is wont to flee from and follow it in turn."

Thus the attractive and repulsive characteristics of magnetism were noted very early, but no mention has been found in ancient literature of the directive property of the geomagnetic field. For many millennia adventurers and merchants went to sea in their fragile craft without any compasses to guide them. They navigated by the sun in the day and the stars at night — a method whose success depends upon some understanding of the geometry of the earth.

Seventeen centuries before Columbus set out to discover a western route to China and to prove that the earth is round, a clever Greek named Eratosthenes demonstrated that the earth is a sphere without traveling more than 500 miles from his home. He measured the angle that the sun's rays made with a true vertical at high noon in Aswan and again in Alexandria, about 500 miles farther north. The angle differed by 7.5 degrees. This would be true, Eratosthenes reasoned, if we were living on the surface of a sphere with a circumference 48 times the distance between Aswan and Alexandria because $48 \times 7.5 = 360$ degrees. By this simple experiment he arrived at an estimate for the size of the earth that was within ten percent of the presently accepted figure.

Pytheas, another Greek who lived almost a century earlier than Eratosthenes, also deduced that the earth must be a sphere. He erected a large sundial near his home in Marseilles and divided the face into units for measurement. He observed the length of the shadow at noon on the day of the summer solstice and from this he computed the angle of the sun's highest point above the horizon. The angle would vary, Pytheas reasoned, with the distance from the equator if the earth were round. As one went farther away from the equator, the seasonal differences in the length of the day would become more and more pronounced. From observations taken at several locations, Pytheas verified these principles and learned to estimate latitude by the length of the days and the length of the shadow on a sundial at noon. Pytheas also recognized the need for finding directions at night. After long months of observing the stars he identified two that lay very close to the true north and did not change their position in the sky. The other stars wheeled in concentric circles around them every twenty-four hours. One of these is now known as the North Star or the pole star. The position on the horizon directly below it marks the true north and the altitude of this star above the horizon gives a measure of latitude.

Pytheas made all these calculations in preparation for a long journey of exploration. Taking his bearings by the sun by day and the north stars by night and keeping a record of the length of the days, Pytheas navigated successfully through the Straits of Gibraltar, along the coast of Spain and France to England. There he landed and spent considerable time traveling through Britain on foot. Returning to his ship, he sailed northward past the Orkney and Shetland Islands. Finally he reached "Thule," which (according to some scholars) was probably Iceland. He even sailed beyond Thule for one day to the "edge of the frozen sea."

This remarkable feat of navigation across the cold and lonely wastes of the northern oceans was achieved sixteen hundred years before the mariner's compass came into general use in the Western world.

The first unambiguous mention of the magnetic needle used to indicate direction was made by a Chinese mathematician and instrument maker, Shen Kua (A.D. 1030–1093), and he refers only to its use on land. The early Chinese compasses were named *ting-nan-ching* — "needle pointing south." A distinguishing mark was attached to the south-seeking needle instead of the north-seeking one, as in the later European compasses.

About 1100 the Chinese writer Chu Yu reported that compasses were used for navigation by "foreign" sailors traveling between Canton and Sumatra. At that time the trade in this area was practically a Moslem monopoly. For this reason it has been assumed that the Arabs were the first to use the earth's magnetic field to find their way at sea, taking advantage of the curious fact that the magnetized needle and the North Star mark out similar paths to the top of the world and define a direction at every point in the Northern Hemisphere. In this way an empirical relationship was established between an ugly, dark stone and one of the smallest, least conspicuous stars among the billions that stud the night sky.

Some of the greatest discoveries in science have resulted from the ability to find a hidden likeness in seemingly disconnected things: to see the march of evolution in the shape of a finch's beak and the color of a tortoise shell, to find a universal force behind the moon's orbit and an apple's fall. But unlike the great clarifying principles of evolution and gravitation, the relationship between the lodestone and the movement of our planet among all the stars of the Milky Way still needs to be illuminated.

❧

Descriptions of the magnetic compass appeared in European literature as early as the end of the twelfth century: "Mariners at sea," said Alexander Neckam, an English cyclopaedist, "when, through cloudy weather in the day which hides the sun, or through the darkness of the night, they lose the knowledge of the quarter of the world to which they are sailing, touch a needle with the magnet, which will turn round till, on its motion ceasing, its point will be directed towards the north."

The floating compass was the earliest type in general use. A needle that had been rubbed with a lodestone was suspended on a stick or straw and floated in water. An imaginative design for the floating compass was described by an Arabian writer in 1282: "They say the captains who navigate the Indian Seas use, instead of the needle and splinter, a sort of fish made out of hollow iron, which, when thrown into water, swims upon the surface, and points out the north and south with its head and tail."

The navigators of medieval times were awed by the wonderful and mysterious ability of magnetized iron to mark out directions all over the earth. They endowed the magnet with mystic powers that they thought must derive from the sun and the stars. The magnetic pole itself was pictured as a glistening black magnetic rock standing in the midst of a whirlpool whose current was so swift that no ship could avoid being sucked into its maw.

At night aboard ship the compass reading was regularly checked by taking "the pilot's blessing." The pilot stood facing the North Star and raised his arm with flattened palm between his eyes so the plane of the palm lined up with the pole star. When the palm was brought straight down on the compass rose, the fingers pointed in the direction of "true" north and the position of the needle was compared with this direction.

Although no one understood why the compass worked, a vast quantity of observations about the earth's magnetic field was collected by the sailing captains who navigated the seven seas. It was discovered that the compass needle does not point exactly to the geographical

North Pole. Furthermore, the difference between the compass needle and true north (the angle of deviation) varies considerably from place to place on the earth's surface.

Christopher Columbus encountered this strange phenomenon as he sailed across the Atlantic in 1492. One week after leaving Teneriffe in the Canary Islands he began to observe a westward shift in the position of his compass needle relative to the North Star. Deviations to the west like this had not been previously reported and caused him great concern. At about this same time the little fleet of three ships entered the Sargasso Sea, an enormous stretch of ocean almost as large as the continental United States, where floating mats of seaweed inhabited by worms and sea slugs and eels covered much of the surface of the water. Winds were very light in these latitudes and the hot sun beat down, accentuating the pungent odor of the weeds. These strange sights and smells filled the sailors with dread. Most of them had never sailed so far out of sight of land before and now the very ocean seemed to be rotting. Wild tales circulated among the men: the weeds will grow up the sides of the ships, creep up the sheets and stays, rooting the ships fast to the ocean bed. They will become decaying hulls prevented from sinking by the seaweed tentacles as the men slowly die of hunger and thirst.

It was at this point, on the evening of September 13, that Columbus first remarked that his compass needle no longer pointed in the direction of the North Star. It pointed about 6 degrees to the northwest. All captains and sailors at that time were acquainted with the fact that the needle does not always point directly north. In European waters it usually assumed an angle slightly east of north. So the total deviation from the position that he and his crew considered normal from their experience was almost 10 degrees. Furthermore, the variation continued to increase over the next few days.

Other officers also noticed the unusual behavior of the compass; their concern quickly spread to the crew, who interpreted the failure of the compass needle to point toward the north as further evidence that they were sailing deeper and deeper into a nightmare world where all familiar objects were contorted into something grotesque and all laws of nature had gone awry. The journal of the trip as preserved by Bartolomé de las Casas reported that "the sailors were frightened and downcast and they would not say why."

Columbus handled this crisis with characteristic decisiveness and imagination. He never revealed his own fear to other members of the expedition. Instead he spoke confidently to them, explaining that the compass needle was still pointing at true north. The deviation, he said, was caused by the changing position of the pole star. It moved slightly over a twenty-four-hour period just like all the other stars in the sky. In the morning, Columbus declared, the sailors would find that the needle again pointed north. He instructed them to check the compass at dawn and, as he had predicted, they found that the needle pointed to the north position on the compass rose.

This report has led to some speculation among scholars. Why did the compass needle point north again in the early morning? Although Columbus's explanation did contain a germ of truth, it was not adequate to account for the readings that were recorded. It is true that the pole star does describe a very small cone in the sky, but the declination caused by this motion is not unique to the Sargasso Sea. If it had been the reason for the large declinations that were observed, the effect would surely have been noted in many other locations.

Another explanation has been suggested. Columbus may have shifted the compass card at night to "correct" for the deviation and allay the fears of his crew. This theory gains credibility when we consider the deception that he practiced to prevent undue concern about the length of the voyage. The journals show that he kept two logs of the ship's track: a true log and one that was altered to make the crews believe that the distance sailed was less than that actually covered. Ironically, it turned out that the falsified figures were more accurate than those calculated by the true log.

We know now that the declination noted by Columbus was a real phenomenon, part of a great eddy of magnetic variation that reached a maximum — perhaps 10 degrees of declination — near the center of the North Atlantic at the end of the fifteenth century. Since then it has moved almost one fourth of the way around the globe. Features like this exist around the planet, like the high and low pressure systems in the atmosphere. But unlike the atmospheric eddies, which usually move rather rapidly from west to east, the magnetic eddies move very slowly from east to west. Old records show that there has been a systematic westward drift amounting to approximately one degree of longitude every five years. About the middle of the sixteenth century there was a point in central Africa where the compass needle pointed directly to the true north — zero declination. This point has moved across the Atlantic Ocean and now occupies a position near the head-waters of the Amazon.

Today there is almost zero declination in the Caribbean area. Attempts to explain the disappearance of ships and planes in the Bermuda Triangle frequently involve the possibility of false compass readings in this region, but actually the Triangle is one of the places in the world where compass variations are at a minimum. The magnetic eddy has moved on and the Caribbean is now a relatively undistorted region.

❧

As the design of the mariner's compass became more sophisticated, the needle was placed on a pivot point, allowing it to move in the vertical as well as in the horizontal direction. If a compass needle is completely free to move in a vertical plane it will assume an angle characteristic of the latitude at which the measurement is taken. In the Northern Hemisphere the north-seeking end of the needle points down from the horizontal. The size of this angle of inclination increases as one travels north until at the pole it points straight down. In the Southern Hemisphere the north-seeking end points up above the horizontal and as one travels south it points higher and higher until at the pole it points straight up. As one moves from the equator toward either the North or South Pole, the intensity of the magnetic field increases. It is almost twice as strong at the poles as it is at the equator.

When considered in this broad aspect the pattern of magnetic intensity and direction seems relatively simple, but when the details are plotted, the pattern becomes almost infinitely complex. Many local variations occur, making strange swirls and burbles in the shape of the magnetic field. Some are caused by iron deposits or discontinuities in the crust of the earth. In many cases, however, the cause is unknown, and in fact the explanation of the whole regular field is still a matter of conjecture.

❧

In 1600 Queen Elizabeth's court physician, William Gilbert, experimented with magnetism. He carved a sphere out of a permanent bar magnet and measured the fields at every point around the sphere. The configuration of these fields matched the regular field of the earth remarkably well. Gilbert theorized that the inside of the earth must contain a large body of permanently magnetic material, making it act like a huge bar magnet. His theory was generally accepted until early in this century, but, ironically, one of Gilbert's own experiments demonstrated the property of magnetism that would eventually disprove the validity of his model. He heated magnetized iron bars to bright red heat and found that they lost their magnetism. According to present theory, the core of the earth does consist largely of iron but its temperature exceeds 4000° F — much hotter than the Curie point, the temperature at which iron loses its magnetism.

Several important phenomena involving magnetism came to light in the nineteenth century. An electric current was found to affect a magnetic needle and, conversely, magnets exert forces on electric circuits. A very effective magnet can be made by inserting an iron core in a coil of wire carrying an electric current. These discoveries led to the invention of the electric motor and the dynamo. They also provided a new way of explaining terrestrial magnetism. According to this theory, electric currents flow within the fluid portion of the core, which is composed mostly of iron. These random currents are maintained by relative motion within the liquid core and their direction is organized by the rotation of the earth. Although this explanation is a reasonable one, many aspects of the model still remain ambiguous. How are the original currents set up? What causes the many fluctuations that have been observed? Why does the magnetic axis not coincide with the axis of rotation? Most physicists and geologists feel confident that the earth's magnetic field can be explained in terms of these physical processes, but a completely satisfactory model still eludes them.

While theoreticians have been working with these problems, more astonishing and perplexing facts about terrestrial magnetism have continued to accumulate. The variations in the geomagnetic field that occur in time have turned out to be even more complex and extraordinary than those that have been mapped out in space. There are cycles that seem to be tied to solar activity. Small differences in intensity (just a fraction of a percent) occur regularly between day and night, between summer and winter, and between years of high and low solar activity. When intense flares of matter erupt from the surface of the sun, a magnetic storm occurs on earth a short time later. Radio communications are severely affected, compasses show "false" readings of several degrees, and brilliant aurora borealis displays are seen in the northern skies. The solar-related variations are readily explained. Changes in the strength of the solar wind (the stream of charged particles flowing outward from the sun) cause induced magnetic fields to be built up within the earth. These transient fields add to or subtract from the regular geomagnetic field. As we probe further back in time, however, we find evidence of much greater changes in the geomagnetic field — changes that have so far defied explanation.

A method of gathering information about ancient geomagnetic fields was suggested by an important discovery in the field of archaeology. When clay objects are fired, a record of the magnetic field direction existing at the time and place of the firing is frozen into the pottery or bricks as they cool below the Curie point. This faint but still detectable magnetism is usually retained throughout the subsequent history of the object and can be used as a clue to the time and place of the object's origin.

When the method was put to use by geologists, they found that almost all rocks contain a magnetic record of their birthplace. Rocks that are relatively rich in iron, such as the igneous rocks that crystallize from molten lava, are most efficient in preserving the magnetic record. The molecules of iron compounds line up like tiny compass needles pointing toward the North Pole. The direction defines an angle of inclination as well as declination; so the magnetization frozen into the rock provides information about the latitude at which the rock was formed and the position of the North Pole relative to the rock at that time. The information is stored in the earth's crust like the magnetic memory elements in a computer.

Studies of magnetism in ancient strata produced some very surprising results. In 1906 a French physicist, Bernard Brunhes, found volcanic rocks that were magnetized in a direction exactly opposite to the direction of the earth's field today, as though North and South poles had exchanged positions. Brunhes concluded that the magnetism of this region of the earth must have been reversed at the time the rock was formed, about 700,000 B.P. (before present). This idea was so fantastic that it was ignored by most scientists.

Other bits of evidence, however, continued to turn up. The Pilanesberg Formation in South Africa, which is more than one billion years old, was found to be magnetized in a direction opposite to the layer

57

directly above it and the layer beneath. On the Columbia Plateau in northwest United States, on the Deccan Plain in India, and on lava flows in Iceland successive layers were also found to alternate between "normal" and "reversed" magnetism. Dating showed that the layers were separated from each other by a few hundred thousand to a million years. Gradually a timetable of geomagnetic reversals was constructed, and the timing of reversals appeared to correspond wherever data were available all around the globe.

As the evidence grew, some physicists and geologists reluctantly accepted the conclusion that North and South magnetic poles had interchanged positions many times during the history of the earth. Others, unable to reconcile this idea with accepted theory, continued to search for alternative explanations of the magnetic reversals recorded in the rocks.

At about this same time another line of scientific research was accumulating information about the present-day magnetic characteristics of the ocean floor. After World War II, when convenient magnetometers were invented, oceanographers took these instruments along with them on their scientific expeditions. A magnetometer could be easily towed behind the boat and readings were automatically recorded on continuous strip charts. When these charts were studied, it became apparent that the strength of the magnetic field varied slightly from place to place as the ship moved across the ocean. As Columbus had discovered variations in the direction of the geomagnetic field, these modern-day mariners found variations in the strength of the field. The variations were not great — only about 1 percent of the total intensity — but they caused strange little ripples on the chart, like the record of a faint murmur on a sound tape. The anomalies were most prominent when the ship passed over the midocean ridges. Most mysterious of all, the pattern of the ripples as the ship approached the ridge was the mirror image of the pattern recorded as it sailed away from the ridge. In the early 1960s oceanographers were particularly interested in these midocean mountain ranges because of the theory that these were the sites of seafloor spreading where new ocean floor was being created. But what could cause the strange symmetrical murmur in the magnetic intensity over these mountain ranges? No one could say. The question baffled oceanographers, but they continued to make the magnetic measurements wherever they went, motivated by the belief in the importance of objective information, whether it could be fitted into a conceptual framework or not. Thus they steadily built up a vast collection of undigested data.

Then in 1963 two young Englishmen, Fred Vine and Drummond Matthews, fitted several pieces of the puzzle together and laid the groundwork for modern plate tectonic theory. The previous year they had taken part in the International Indian Ocean Expedition, making a detailed magnetic survey over the Carlsberg Ridge. It occurred to them that the mirror-image magnetic anomaly patterns could be explained on the basis of ocean-floor spreading and periodic reversals in the earth's magnetic field. As hot magma rises at the center of an oceanic ridge, it cools and is magnetized in the direction of the field at that moment in time. Then, as spreading occurs, bands of this solidified lava move away from the ridge, symmetrically to the right and left. When the earth's field reverses, the lava which is crystallizing on the ridge at that time becomes magnetized in the reverse direction, and bands of this rock also moves out on either side of the ridge. So alternating bands of normal and reverse magnetization drift away from the ridge center. A magnetometer pulled across the surface of the ocean above the ridge records the presence of these parallel bands — the normally magnetized bands adding to the field of the earth and the reversed ones subtracting from it. The width of each band is a function of the speed of ocean-floor spreading and the length of time between reversals. The chronology of the magnetic reversals, which had previously been documented in lava flows on land, provided the time scale for the reversals.

When Vine and Matthews first proposed this idea, it met with considerable skepticism. In 1963 Vine gave a talk explaining the theory to a group of geologists and oceanographers at Lamont-Doherty Geological Observatory. The silence that greeted the end of his presentation told him that he had not convinced his audience. With sudden inspiration, he suggested that they bring out some of their charts of magnetic readings over ocean ridges and he would show how he could identify the anomalies on a worldwide basis. So challenged, they delved into their store of records and were deeply impressed when Vine was able to match the little ripples on the traces with those taken in other parts of the world. Overnight the collection of undigested data became a treasurehouse of information. The mysterious magnetic murmurs were telling a wonderful story about the rate of seafloor spreading, the creation of ocean basins, and the movements of pieces of earth crust around the world. It is pecularly fitting that the Rosetta Stone of the earth's magnetic history lies at the bottom of the sea, thousands of fathoms beneath the wave-tossed surface where medieval mariners first learned to find their way in the dark of night using a little sliver of magnetized iron floating on a straw.

❧

After the key to the magnetic language had been discovered, the story unfolded rapidly. As the chronology of the reversals was worked out, it became evident that periods of normal and reversed magnetism have been almost equally represented throughout the time, which can be dated with reasonable accuracy. The transitions took place almost instantaneously on the geologic time scale. Typically, the intensity of the field declined over a period of about 10,000 years to about 60 to 80 percent of its former value. Then during the next thousand years or so it decreased rapidly toward zero. The flipover occurred suddenly and the field built up at about the same rate in the other direction. Throughout the last eighty million years, the average interval between normal and reversed epochs has been about half a million years. However, several very short reversal events are also recorded in the rocks. These episodes — or excursions, as they are called — lasted for a few thousand years. The most recent event of this type took place during the last ice age and must have been witnessed by Stone Age man. Except for this brief event, the earth's magnetic field has been in its present orientation for the last 700,000 years. This is an abnormally long epoch; only 15 percent of the normal epochs were of longer duration. The evidence suggests that the earth is overdue for a change.

Since 1835 the earth's magnetic field has decreased at a fairly uniform rate amounting to a total of about 6 percent. If the rate of decrease continues unchanged, the magnetic field will be reduced to about a tenth of its normal value in less than 2,000 years. Perhaps we are witnessing the beginning of another reversal.

There is, however, another possibility. The record shows that the strength of the magnetic field has passed periodically through shorter cyclic changes without going through a reversal. The intensity has varied by more than a factor of two, following a regular pattern. The maximum of the present cycle occurred around A.D. 100, and most earth scientists believe that the declining field we are experiencing today is part of this shorter cycle. But no one is really sure. The reasons for the shorter and longer cycles are equally obscure.

~~

What would the world be like if the magnetic field were reduced to a fraction of its present value? What if compass needles on a million ships and planes began to swing idly, pivoting back and forth, and finally came to rest with the north-seeking needle pointing to the south?

In order to judge the impact of such a strange event we can look back to the last well-documented reversal in the geologic record. The Gothenburg Excursion, as it is called because evidence of it was first found near the town of Gothenburg in Nebraska, occurred about 13,500 B.P. and lasted for one or two thousand years. A number of dramatic events happened at approximately that same time. Enormous glaciers that had covered most of North America and northern Europe during the height of the last ice age (approximately 18,000 years ago) had been slowly retreating for thousands of years. The climate, which had been excessively dry, was becoming more humid. Millions of tons of water previously locked up in glaciers were beginning to flow back into the oceans and evaporate into the atmosphere. Rain was becoming more frequent and vegetation was slowly encroaching on the edges of the tremendous sand deserts that covered much more of the earth's surface than they do today. Winds were moderating a little, although they were still strong enough to carry inch-size angular stones, battering them with destructive violence against the tundras at the edge of the ice field.

The site of present-day Manhattan Island lay many miles from the sea and had been covered by glaciers several thousand feet thick. But now it was beginning to emerge from its tomb of ice. The Dordogne Valley in France had never been glaciated at the height of the last ice age; it was a bare, forbidding tundra on the southern edge of the ice fields. Cro-Magnon men had found shelter against the rigors of the climate in large limestone caves. On the walls they painted portraits of the animals that shared this rugged environment with them: mammoths, woolly rhinoceros, bison, and reindeer. In the year 14,000 B.P. conditions were improving in the Dordogne Valley also. Vegetation had begun to clothe the barren slopes along the banks of the Loire.

Suddenly, about 13,500 B.P. the glaciers stopped retreating and reoccupied large areas of their former territory. This event is well documented in the United States, Canada, northern Europe, and New Zealand. Within a period of a thousand years a new volume of ice formed, equivalent to twenty times the combined size of the world's present mountain glaciers. Sea levels dropped again as much as thirty-five feet. The general cooling appears to have persisted with some fluctuations for nearly two thousand years. Then around 11,600 B.P. warmer weather returned, causing a major flooding of low-lying coastal areas and giving rise, perhaps, to the stories of great floods that are part of the oral tradition of almost all cultures.

These events coincided closely with the onset and termination of the Gothenburg Excursion. Were these changes directly related to the reversal of the field?

There is some theoretical support for the hypothesis that a magnetic reversal could cause a pulse of cold climate in the temperate latitudes. When the strength of the magnetic field is greatly reduced, cosmic radiation, which pervades all space and which is normally deflected away from the earth by the geomagnetic field, is able to enter the atmosphere, causing a great increase in ionization. Ions act as nuclei around which cloud droplets form; therefore many more clouds would form in the upper atmosphere, screening out some of the sun's rays. Cooling would occur and this trend would be reinforced as more ice formed because more of the sun's heat would be reflected rather than absorbed by the earth. Drought conditions would become prevalent as glaciers locked up more of the planet's water supply. So a cold, dry, cloudy period could be expected during a magnetic reversal and these conditions are generally unfavorable for life.

In fact, some scientists believe that magnetic reversals have had a detrimental effect on living things. On several occasions massive disappearances of certain one-celled marine organisms have occurred close to the time of major changes in the geomagnetic field (see Chapter 12). Furthermore, the great extinctions of life that occurred 225 and 65 million years ago coincided with periods when long intervals of normal polarity were followed by polarity fluctuations. But other scientists doubt that magnetic reversals, even if accompanied by pulses of cold climate, could have caused these crises in the history of life. Most organisms, they point out, should have been able to adapt to the temperature changes, which would have occurred gradually over a period of at least several thousand years. And the instances when extinctions occurred at approximately the same time as reversals in geologic history may have been purely coincidental. So the theory that magnetic reversals caused major die-offs remains very controversial.

In the meantime, evidence is mounting that magnetic fields exert an influence on living things in other, more subtle ways. Experiments have shown that primitive creatures, such as the mud snail and planaria, can distinguish between different magnetic intensities and adjust their orientation to maintain a certain relationship with the magnetic field. *Volvox*, a tiny motile alga similar to some of the earliest forms of life, alters the path of its movement in response to a changing magnetic field. According to biologist Frank Brown, who has conducted many of these studies, the fact that *Volvox* can use magnetic fields in spatial orientation may indicate that the ability to respond to magnetic fields evolved in organisms soon after their inception on earth and that this trait was passed on to higher organisms like the birds.

It has long been suspected that birds use some biologic equivalent of the mariner's compass to guide their migrations, but many experiments have been performed to test this hypothesis and the results are somewhat ambiguous. Birds seem to use a number of different directional clues just as sailors do in navigating across the sea. The magnetic field is one of these methods, but no one is certain how birds would respond if the magnetic field were reversed or reduced to zero.

It is possible that human behavior is also affected by magnetic fields. An interesting statistical relationship was discovered between the number of admissions to psychiatric hospitals and changes in the geomagnetic field. Records from seven hospitals in the same area over a four-year period were compared with magnetic readings taken by the Coast and Geodetic Survey. They showed a correlation of a very significant magnitude. Although correlation in itself does not prove a causal relationship, the results of the study suggest some intriguing possibilities.

It would not be surprising to discover that all forms of life respond to changes in magnetic fields. The further scientists probe into the nature of matter, the more universal and basic the property of magnetism appears to be. From the great spinning galaxies to the tiniest "fundamental" particles that make up the atom, each individual unit of matter has its own characteristic magnetic properties.

As these facts have unfolded and as the almost infinite complexity of the geomagnetic field both in time and in space has been revealed, some earth scientists have begun to believe that magnetism is part of a still unexplained cosmic process. The cause of terrestrial magnetism may not lie within the earth itself at all but in its relationship to the sun and the galaxy and the rest of the universe. At first glance this suggestion seems to represent a retreat from science to mysticism — a return to the thought of the ancient philosophers that the magnet derives its power from the sun and the stars. But science deals routinely with mysteries, with unsolved puzzles, with the unexplained. One day the hidden relationship will be revealed — the unifying principle that binds together the flight of the homing pigeon and the compass rose, the lodestone and the pole star, the nebula and the quark.

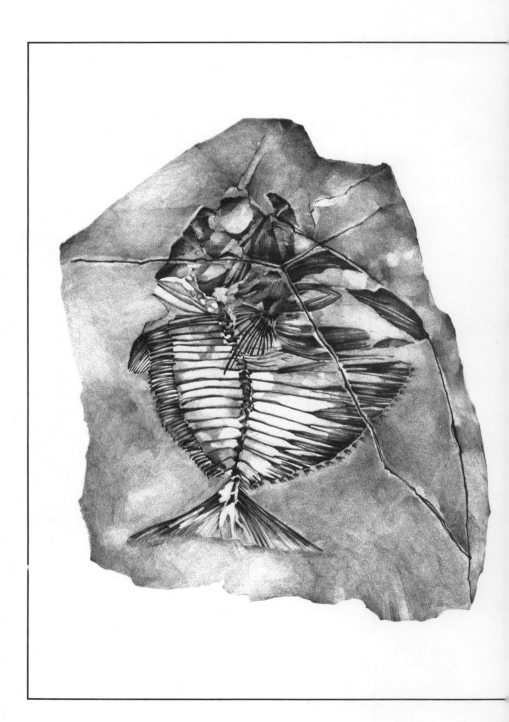

4

Pangaea and Other Puzzles

"I can't believe that," said Alice.

"Can't you?" the [White] Queen said, in a pitying tone. "Try again: draw a long breath, and shut your eyes."

Alice laughed. "There's no use trying," she said: "One *can't* believe impossible things."

"I daresay you haven't had much practice," said the Queen. "When I was your age, I always did it for half-an-hour a day. Why, sometimes I've believed as many as six impossible things before breakfast."

— LEWIS CARROLL, *Through the Looking Glass*

Like Alice, most people have trouble believing impossible things and scientists are no exception. In fact, scientists are sometimes more reluctant than the average to accept enthusiastically a radical new idea. It is their business to explain how things that appear to be impossible do in fact occur. The feat requires imagination and independence of thought, attributes that are rare in any calling. When some gifted theoretician does suggest an important departure from conventional theory to account for new discoveries, the idea is frequently rejected by the rest of the scientific community. Attempts are usually made to fit the new facts into the old pattern of thought, and many years may go by before enough evidence accumulates to necessitate a major overhaul in accepted theory. Then the rejected idea may be resurrected and adopted as the "new" paradigm.

The history of science is replete with examples of brilliant ideas that were not appreciated in their own time. Centuries passed before the Copernican theory replaced the clumsy Ptolemaic model of the universe. Mendel's revolutionary genetic research was ignored for fifty years before it was rediscovered and recognized as an important concept, providing a mechanism for Darwin's theory of natural selection. Hooke's persuasive arguments for the extinction of species were not generally accepted until almost two centuries later.

It is interesting to speculate how many man-hours of fruitless scientific work would have been saved if the new concepts had been more carefully evaluated when they were first proposed. Conservatism of thought is a handicap in scientific research; the truth usually proves to be more astonishing than the most radical hypothesis. Perhaps all young scientists should be assigned the exercise recommended by the White Queen — to practice believing impossible things for half-an-hour before breakfast.

A recent example of scientific conservatism — one that occurred in our own century — was the rejection of Alfred Wegener's theory of continental drift. Long before Wegener's time prominent thinkers had noted a striking similarity in the shapes of Africa and South America. Sir Francis Bacon remarked in his book *Novum Organum* (1620) that such a close correspondence could hardly be accidental. In 1858 Antonio Snider-Pelligrini published a map showing the continents on either side of the Atlantic fitted together like pieces of a jigsaw puzzle, and one of his contemporaries, J. H. Pepper, used this diagram to explain the finding of similar fossil plants in the coal fields of Europe and North America. The idea that the continents had split apart, however, was far ahead of its time and was not taken seriously either by fellow scientists or by the general public.

By the end of the nineteenth century scientists had discovered many examples of identical fossil species in South America, Africa, India, Australia, and Antarctica. A distinctive group of plants grew on most of these southern continents 300 million years ago. The most prominent member of this community was a plant whose leaves (named *Glossopteris*) were long, slender, and oval, shaped like pigeon feathers. Although the seeds of this plant were too heavy to be carried on the wind, the fossil leaves have been found in Antarctica as well as in

India, more than seven thousand miles away. All of the plants in the *Glossopteris* group were characteristic of vegetation adapted to a cool seasonal environment. Other geologic evidence supported the conclusion that large regions of India, Africa, South America, Australia, and Antarctica were heavily glaciated about 300 to 275 million years ago. On all five southern continents unmistakable signs of glacial deposits have been found: long striations scratched into ancient rocks by boulders embedded in slowly moving masses of ice.

At this same time large areas of the northern continents were clothed with luxuriant forests of tropical vegetation. The patterns of these ancient forests are quite distinct from the patterns of vegetation that characterized the forests of South Africa, India, South America, Australia, and Antarctica. Here lycopsid trees grew tall to catch the sunlight with their topknots of palm-shaped foliage. Dragonflies as large as mourning doves sailed on gossamer wings above marshes where ferns dipped graceful fronds into the stagnant water. Leaves filled the forest spaces — pale green, yellow green, velvety moss green — all variations on the dominant color of life. But no trillium or lady slippers lighted the deep shadows. No orchids hung from the jointed sphenopsid trees. Flowers were still a hundred million years away. And no birds sang.

～

Remains of several animal species have also been found widely distributed on southern continents. Fossils of *Mesosaurus*, a snaggle-toothed reptile about eighteen inches long, have been uncovered in both Brazil and South Africa. A shallow-water creature that lived about 250 million years ago, *Mesosaurus* could not have swum very far — certainly not across the ocean that now exists between these two lands. A larger land reptile named *Lystrosaurus* inhabited Africa, India, and Antarctica about 200 million years ago. These fossil finds are the most spectacular examples of a large number of identical or very similar fossil species found in places now separated by deep oceans.

Many of these facts were known by the early years of the twentieth century. In order to explain them naturalists suggested that land connections must have existed between the continents. They believed that the continents themselves had remained fixed in their present positions and the intervening land had sunk beneath the surface of the sea.

Explanations based on theories of continental movement were

suggested independently by two Americans, Frank Taylor and Howard B. Baker, in 1908. But it was a German scientist who worked out the most detailed and extensively documented version of this hypothesis.

～∾

Alfred Wegener was born in Berlin in 1880. He was trained as a meteorologist and astronomer at the universities of Heidelberg and Innsbruck. In 1906 he was invited to join a Danish scientific expedition to Greenland to gather information on polar air masses. This trip proved to be a very arduous experience; the members of the team fought a constant battle against cold and privation in a country still held in the grip of the Ice Age. The land was permanently frozen, bare and arid as a desert — continually exposed to bitter winds. The leader and two other men died during an overextended dogsled trip into the northern interior. But Wegener returned safely, his enthusiasm for Arctic exploration still bright, his memory stoked with vivid impressions of the interesting geology of the world's largest island. Back in Germany he was offered a teaching position in meteorology and astronomy at the University of Marburg, a post he held until his second trip to Greenland in 1912. During this time the theory that was to make him famous was germinating in his mind. He first conceived the idea in 1910, when he noted the remarkable degree of correspondence between the shapes of the continental coastlines on either side of the Atlantic. The possibility of continental drift occurred to him, but he rejected it as being improbable. Then in 1911 his interest was reawakened when he read about the similarity of fossil remains found in South America and Africa. This fact was being used to support the theory of a land connection between Brazil and Africa. Far more likely, Wegener thought, that the two continents had actually been one. He expounded this hypothesis in a lecture delivered in 1912 and later that year published several papers on the subject. In 1915 an elaborated version of his theory appeared in book form, *The Origin of Continents and Oceans.*

About 200 million years ago, Wegener said, there existed a single continent that he named "Pangaea." Antarctica, Australia, India, Madagascar, Europe, Africa, Asia, and the Americas were all nested together in this single landmass. The rest of the world was ocean — the proto-Pacific. Then Pangaea began to rupture and drift apart, opening up the Indian and Atlantic oceans as the continents separated.

Wegener supported his hypothesis with an impressive collection of facts: the geometric fit of continental margins, corresponding rock successions, similar ancient climatic conditions, and identical life forms on continents now widely separated but contiguous in his reconstructed Pangaea. All in all, his scenario was remarkably close to accepted theory today.

Although the book attracted considerable attention and stirred up lively discussion among earth scientists, most of the reaction was negative. Geophysicists declared the theory preposterous. There was no known mechanism that could account for such extraordinary movements of whole continents. The lighter-density granitic crust could not have moved through the denser basaltic crust of the ocean floor. Unfortunately, Wegener himself was not able to propose any plausible physical explanation. The driving force, he suggested, might be due to the rotation of the planet or to the gravitational pull of the sun and the moon, but geophysicists were quick to disprove these hypotheses, treating them with sarcastic scorn. Wegener's idea did gain a few supporters, especially on the southern continents, where the evidence was most strikingly demonstrated. These scientists argued that the impressive evidence for drift should not be ignored just because geophysicists were unable to find a physical explanation for it. But in the United States and much of Europe geologists did not consider the evidence impressive. The theory of continental drift was either ignored or treated as a crackpot hypothesis. The fact that Wegener was a meteorologist and an astronomer — an outsider in the field of geology — did not help his cause.

In Germany Wegener received some personal recognition although his theories were not viewed with favor. A professorship in meteorology and geophysics was created for him at the University of Graz in 1924. And he led two more expeditions to Greenland in 1929–1931. On these trips Wegener hoped to find more definite proof for his theory of continental drift. Using his training as an astronomer, he made measurements of distances on the earth's surface by triangulation from stellar positions. He also experimented with measurements of radio time transmission across oceans to calculate distance between continents. By accumulating these data over a period of years he hoped to show that Greenland is actually drifting westward. Unfortunately, his experiments were doomed to disappointment. Wegener was dealing with processes so slow in the human time scale that, given the methods at his disposal, he could not have detected any movement between continents even in a normal lifetime. And he had only one year more to live.

Wegener pioneered more successfully another technique on the Greenland expeditions: an echo-sounding device for measuring the thickness of the ice. This instrument was the prototype for sonar, which later became one of the important tools for studying the ocean floor and by an ironic twist of fate, it produced information about the mid-ocean ridges that helped to verify the theory of continental movement many years later.

The physical conditions on the 1929–1931 expeditions were even more rigorous than those experienced on the earlier explorations. A station was established in the middle of Greenland. Called Eismitte, "the middle of the ice," it was nearly 250 miles from both coasts. The supplying of this outpost strained the strength and resources of the research team to the breaking point.

In October 1930 a message reached the coastal station that the two men at Eismitte could not survive through the long dark winter unless they received more food and fuel. Wegener set forth with two companions and additional supplies on two dogsleds. They reached Eismitte on October 27. Temperatures there had already fallen to 60° below zero F, with wind speeds gusting up to 70 miles an hour. Day by day the sunlight was draining from the sky, leaving longer and longer hours of unearthly twilight.

Wegener celebrated his fiftieth birthday at Eismitte and then set off with one Eskimo guide on the long journey back to the coast. We can imagine the fast-gathering winter darkness, the biting winds, the seemingly endless trek on skis across ice-encrusted snowdrifts whose edges were sharp as shards of broken glass. Wegener did not reach the expedition's home base. His body was found the next spring about halfway between Eismitte and the coast of Greenland, where his Eskimo guide had buried him in a shallow depression in the snow. The guide himself and the dogsled were never found.

❧

Wegener had failed to convince most of the scientific world that continents have not remained fixed but rather have drifted slowly around the planet. If he had lived to be eighty-seven years old, he would have seen his ideas rediscovered and himself honored as a brilliant scientist.

In the meantime geologists got on with the difficult work of fitting the known facts into the accepted theories of the time. As paleontologists discovered more and more evidence that the same species had inhabited continents now separated by thousands of miles of open ocean, it became necessary to rely heavily on the existence of land connections that later disappeared beneath the sea. Since the rocks of such continental fragments would have been granitic rather than basaltic in composition, they could be identified by sampling of the ocean bottom. But when these experiments were carried out, not a single trace of drowned land was found in the expected places except in the Bering Strait and the Scotia Arc between South America and Antarctica.

As evidence against this explanation mounted, the theoretical land connections were continuously reduced in size. They became narrow corridors or "bridges." More reliance was placed on island hopping, rafting, and wind dispersion. Of course, seeds and birds can be carried on the wind and small species may be transported on rafts of vegetation, thus accounting for the worldwide distribution of some species. But in other cases, such as the large dinosaurs that have been found on every continent except Antarctica, these explanations were stretched beyond plausibility.

During the first half of the twentieth century studies of the magnetic field recorded in ancient rocks turned up some relevant information. As we have seen, the direction of the magnetic pole at the time of the rocks' formation can be inferred from the magnetism preserved in the stone. When these ancient pole positions were plotted, it appeared that the poles and the continents had moved in relation to one another throughout at least several hundred million years. This fact could be readily explained if the landmasses had migrated throughout that same period of time as Wegener postulated.

In the 1960s the discovery of magnetic stripes on the ocean floor and the dating of the rocks along the midocean ridges provided dramatic confirmation of seafloor spreading. From this information it was even possible to estimate the rate at which the Atlantic Ocean had opened up, separating the continents of Africa and South America, Europe and North America. The time taken to create the Atlantic Ocean turned out to be approximately the same length of time that Wegener had postulated — about 200 million years.

Perhaps none of these facts would have been sufficient to cause a revolution in geology were it not for a new understanding of earth structure that provided a physical explanation of crustal movement. The study of earthquakes and the manner in which the shock waves travel through the earth had provided important information about the insides of the planet. Just as a doctor thumping a patient's chest can detect areas of abnormal density, so the reflection and absorption of earthquake waves yield information about the interior of the earth.

It was found that the planet is layered like an onion. First there is the thin exterior crust, five to twenty-five miles thick; then the mantle extends some 1,800 miles below the crust. Beneath the mantle lies the liquid portion of the core — about 1,380 miles thick; and finally a solid sphere occupies the center. Each successive layer is under increasing pressure from the weight of the overlying material. At the very center pressures are 3.5 million atmospheres and temperatures are nearly 4500° F. The composition of the earth also changes with depth below the surface. The crust contains many light elements — hydrogen, oxygen, silicon, calcium, sulfur — as well as some amounts of iron, magnesium, and other heavy elements. The mantle has a smaller share of the light elements and a larger portion of heavy elements. The core is almost entirely composed of iron with a small amount of nickel and possibly a trace of silicon or sulfur. In the upper mantle there is a layer of semisoft material where the rocks are near the melting point. This magma moves plastically in response to steady pressure, like tar or Silly Putty. Geologists theorize that within this layer, known as the astheno-sphere, heat causes convection currents. As the material expands, it rises slowly to the top like a bubble in a kettle of thick soup. It presses against the underside of the brittle layer (the lithosphere, which contains the crust and the solid part of the upper mantle) and causes fractures wherever it encounters a weak point. Then as the current flows out to either side it carries the lithosphere along with it and the fracture widens. Soft, semiliquid matter wells up to fill the crack and in this way the great midocean ridges are formed. As the fracture opens, the sea-floor spreads. The stripes of normal and reversed magnetism preserved in the new rocks provide the key for calculating the rate of spreading.

This convection theory provided the long-sought mechanism that would make the impossible idea of drifting continents possible, after all. But one troublesome problem remained to be solved before the explanation could be considered complete. Since the area of earth on the ridges is increasing, either the planet must be expanding or earth surface must be destroyed somewhere else. The latter possibility can be accommodated within the convection model. The hot rising magma causes upward pressure, then as it moves laterally beneath the lithosphere it cools, contracts, and begins to sink again, causing a downward pull on the brittle layer directly above it. The lithosphere at this point cracks and is dragged down into the mantle, where it is heated, melted, and re-cycled. According to this model, the descending limbs of convection currents should create deep depressions and, in fact, a number of major fractures have been discovered on the ocean floor. They lie near regions of frequent earthquake activity and are especially characterized by very deep quakes. Heat flow from these trenches is not abnormally high as it is at the spreading fractures where hot magma is boiling up. In all these ways the trenches seem to fit the description of subduction zones as required by the theory, and therefore they are believed to be the places where ocean crust is being sucked back into the mantle. The lithosphere, fractured in the regions of upwelling and down-flowing currents, is divided into rigid plates, each moving as a unit across the surface.

❧

When these facts and ideas came together in the years between 1963 and 1967, a great revolution occurred in geology. Within the space of a few years most scientists in the field became enthusiastic about this dynamic view of the earth. Plate tectonics became the new paradigm.

Taking the ridges and the trenches as the boundaries of the pieces of earth crust, geologists identified six major plates and several minor ones (see Figure 2). Almost all of these active regions are on the seafloor — a fact that is not surprising because the planet's crust is much thinner there than it is on the continents. Fracturing can occur more easily on the thin ocean crust.

Although continental crust is thicker than oceanic crust, it is composed of lighter and frothier stuff. So whenever a plate carrying continental crust collides with an oceanic plate, the former rides up over the latter and the oceanic lithosphere is subducted. This is the explanation of the fact that the ocean floor is very new while the rocks of the continents represent all ages; some are almost four billion years old.

Many of the predictions based on the plate tectonic model fit remarkably well with the known facts. There remain some objections, however, that have not been clearly resolved. For example, no one has demonstrated that crust is actually disappearing in the trenches or that the amount of crust subducted is equal to the amount forming along the rift zones. Ridges and trenches are quite unevenly distributed around the world; so it would seem reasonable to expect that areas of ocean floor would be missed — not cleanly swept away by the subduction process. But no areas of seafloor older than 180 million years have been found. As more of the ocean floor is sampled, perhaps some very ancient oceanic crust will be discovered, thus providing strong support for plate tectonic theory.

In the meantime there are many reasons to be enthusiastic about the theory. The creation of mountain ranges and island arcs, the distribution of volcanoes and earthquakes, can all be accommodated within this model of the earth. Since granitic crust resists subduction, wherever two continental plates collide, massive buckling, crumpling, and folding of earth material occur. Such collisions are believed to be responsible for the formation of many great mountain ranges like the Himalayas and the Alps.

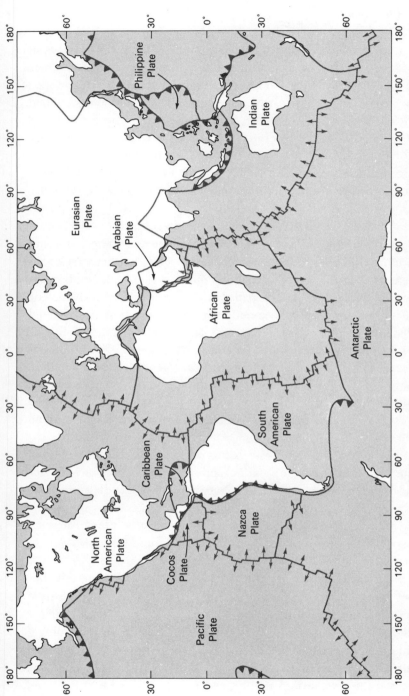

FIGURE 2   Twelve plates and their motions. Triangles along boundaries indicate direction of underthrusting where a downgoing slab can be identified by the occurrence of intermediate or deep focus earthquakes. Small arrows on ridge boundaries indicate approximate direction of relative motion. (From D. Forsyth and S. Uyeda, "On the Relative Importance of the Driving Forces of Plate Motion," Geophysics Journal 43, p. 163, 1975)

The trenches lie adjacent to volcanic mountains and island arcs in belts of frequent earthquake activity (see Figures 4 and 5). The entire Pacific Ocean (except the very southernmost portion) is surrounded by these regions of active earth movement, which have been dubbed "The Ring of Fire." The Pacific, according to this theory, is being slowly reduced in size, the ocean floor sucked into the great trenches. Some of this substance is heated by friction and forced back up in volcanic eruptions creating mountain systems like the Andes and island arcs paralleling the lines of the trenches.

One of the most extraordinary places on earth is a little corner known as the Afar Triangle in East Africa, where three spreading rift zones meet and lithosphere that is essentially ocean floor has been raised up above sea level. In this region, which lies northwest of Djibouti, bordering the Red Sea, the effects of rifting are dramatically displayed. For hundreds of miles the earth is covered with black lava flows and littered with basaltic boulders that are perforated with holes made by air bubbles as molten lava erupted under great pressure. These porous rocks are thickly strewn across the land as though a hundred million enormous sponges had been dipped in dark ink and left to harden under the scorching tropical sun. In some regions they are black, in others a deep rust red, extending from horizon to horizon like a landscape on Mars.

One of the few "roads" out across this desolate land winds its tortuous way over loose rock piles, across the flanks of ancient volcanoes, and down steep escarpments between Ghoubbet Bay and Lake Asal. Important earth movements have taken place here just in the last few years. In November of 1978 a new volcano erupted. For three days and nights the whole region trembled. During one twenty-four-hour period 2,000 earth tremors were recorded.

When I visited the site in 1981, the new volcano was quiescent but cones of many earlier eruptions were scattered helter-skelter across the countryside. The road passed close to the edge of a rift that plunged precipitously into a narrow gorge thousands of feet below. Strata of light and dark colored rock had been laid bare as though a sharp knife had cut a wedge out of a many-layered torte. This rift is only one of many fissures splitting the earth between Ghoubbet Bay and Lake Asal.

On every side there are indications that this land lay beneath the surface of the sea not so very long ago. Almost the entire crust is basaltic in composition, like the ocean floor. There are mounds of pillow lavas — that distinctive formation made when lava erupts beneath the sea and solidifies quickly on contact with cold water. Some rocks glisten with bits of volcanic glass produced by extremely rapid cooling. And across many acres the lower portions of the large boulders are encased in a layer of limestone. This crust reaches to the same height on all the rocks, apparently delineating the high-water mark of an ancient sea. Hillocks of porous white limestone stand out like pyramids, startlingly bright against the black and rust-red color of the lava fields. Perhaps the most curious sight is a field of jagged, pale ivory-colored spikes and tiny nodules — a deposit made by bubbling hydrothermal vents.

In a number of places evaporite deposits are forming as lakes and pools dry up in the oppressive desert heat. These bodies of water receive no river influx; they are fed by hot, sulfurous, salty springs, like the hydrothermal vents on the seafloor. Some of these are scalding hot where they emerge from the earth. And yet, as in the similar springs in the midocean rifts, certain forms of life proliferate there. Algae make a thick carpet of green on the stones over which these waters flow.

The Afar Triangle lies at the point where the East African Rift Valley meets the spreading zone between the Gulf of Aden and Arabia and the rift that runs down the middle of the Red Sea (see Figure 2). Coring and profiling in the sea have shown this long, narrow channel of water to be a new ocean in the making. A smooth continental shelf extends into the sea from each shore. Then there is a sudden drop-off into a deep axial trough that runs the entire length of the sea and is itself split down the center by another rift that plunges to great depths — nearly 10,000 feet at the deepest penetration. The continental shelf areas are topped with a thick bed of sediment, but below this there is a layer of salt and gypsum deposits that were formed about twenty to thirty million years ago. The layers of sediment and evaporite are completely absent in the axial valley where distinctive lava formations indicate seafloor spreading within the last two million years. A series of cores and photographs taken in the axial trough have revealed large pillow lavas, volcanic fissures, and deep pools of hot brine. Temperatures range up to 140° F, and the waters are extremely saline. Salt concentrations are as high as any in the world: 300 parts of salt to 1,000 parts of water have been recorded there as compared with 35 parts salt for average seawater.

The hot brine pools of the Red Sea are also rich in heavy metals, such as copper, lead, zinc, iron, vanadium, and silver. These metals have been carried up from the inside of the earth in magma and deposited by chemical action with the seawater. The process is similar to that which has been observed in the hot thermal vents on the rift off the Gulf of California. In one deep area it is estimated that metal deposits worth more than two billion dollars have accumulated. But unfortunately the cost of recovering them from one or two miles below sea level is so great that it is not an economical venture at the present time.

The shape and age of the layers beneath the surface of the Red Sea suggest that the formation of this body of water took place in two stages. The first occurred twenty to thirty million years ago as the result of rifting between the African and Arabian plates. Then spreading stopped. A long period of dry climate produced extensive evaporation followed by many years of rainier conditions when sediments from the land were deposited on the shelf. Finally a new stage of rifting began about two million years ago and is still continuing today.

It is interesting to note that the oldest hominid fossils, dated between 3 and 3.5 million years old, have been found in the Afar Triangle. Was there some reason why our remote ancestors were particularly attracted to this piece of earth crust? Perhaps during the rainy period before the last rifting, this land, which is so bare and forbidding today, was a favorable environment, a warm, protected valley containing many lakes and streams. It is also possible, however, that those ancient forebears of ours occupied a wider territory which included the Afar Triangle, and in this one place the special conditions leading to the preservation of their bones were uniquely realized during one of the periods when the crust of the earth subsided and the waters of the Red Sea inundated the low land. The anthropologists who have studied the fossil remains of the tiny hominid Lucy discovered here were impressed by the fact that the skeleton had been protected from the attack of predators, preserved from dissolution. "Her carcass remained inviolate, slowly covered by sand or mud, buried deeper and deeper, the sand hardening into rock under the weight of subsequent depositions. She had lain silently in her adamantine grave for millennium after millennium until [the land was raised up again and] the rains brought the skeleton to light."

The rising and falling of this land, the rending and cracking of the earth's crust, signal the beginning of a major creative work, the making of a new ocean. The magnetic stripes on the Red Sea floor reveal its rate of spreading — about half an inch a year. At this rate, in 200 million years the Red Sea will be as large as the Atlantic Ocean is today. The Atlantic took approximately that same length of time to form and we can imagine that in the early stages of rifting the Atlantic looked very much like the Red Sea does now.

∽

Lake Baikal in Siberia is another long, narrow, and extremely deep body of water that is believed to be an embryonic ocean. Its basin is more than a mile deep and contains more water than all the Great Lakes combined. Below the water, soft sediment deposits extend for another four miles before bedrock is reached. The shape and age of the steep walls reveal the fact that this basin was formed by a sharp fracture in the earth's crust occurring about twenty-five million years ago.

As an ocean in the making, Lake Baikal is an even more amazing phenomenon than the Red Sea. Baikal lies a thousand miles from any ocean. It is filled with cool, fresh water — a far cry from the hot brines of the Red Sea. The plate movements that formed the lake are not as readily apparent as they are in the case of the African and Arabian plates. But Lake Baikal does lie on a fault zone that runs across Mongolia and Siberia to the Arctic Ocean. Thirty major earthquakes have been recorded at Baikal over the last two centuries. In 1861 seventy-seven square miles of land collapsed and fell into the lake. Thirteen hundred people were killed and a large new bay was added to the lake.

Because of its remote and inaccessible location we know less about the geology of this lake than we do about the Red Sea. However, two dives in submersibles to the bottom in 1977 discovered the presence of hot springs and active volcanism beneath the soft sediments — almost five miles below the cool surface of this unusual body of water. There is every reason to believe that Lake Baikal will gradually grow wider as the great continent of Asia splits apart. The sites of Irkutsk and

Ulan Ude will move steadily away from each other, riding on separate landmasses as Lake Baikal becomes the Siberian Sea.

This lake provides a wonderful fresh-water environment for large numbers of aquatic species. It is rich in oxygen, startlingly clear, perhaps the cleanest large body of fresh water on earth. Because it has been an isolated aquatic environment for many eons a great many of the lake's organisms have evolved in their own special ways to the point that they are unique endemic species. Twelve thousand of the lake's 17,000 indigenous life forms are not found anywhere else in the world. The most spectacular inhabitants of the lake are the fresh-water seals called *nerpy* in Russian. They are related to the Arctic seals from which they must be descended, but their fur is a lovely silvery gray instead of brown. Their biology and bone structure differ from the Arctic seal's enough to make them a distinct species. Sixty thousand of these engaging creatures play in the cold waters and loll on the rocky lakeshore in the sun.

Other favorite endemic species are Baikal sturgeon and *omul*, a salmonlike whitefish that is considered to be a great delicacy. One of the Baikal species, *golomyanka*, is perhaps the oddest fish in the world. In order to protect itself against the cold waters, this fish has evolved a very efficient way of incorporating fat into its tissues. Its body contains such a high percentage of fat that it is actually translucent. But if the fish is removed from the water and placed in sunlight, the fat literally melts away and the body is reduced to skin and bones. This fish is also unusual in its manner of reproduction. It gives birth to larvae instead of eggs and usually dies in the act. The female's body splits asunder to release the young.

The number of unique species found at Lake Baikal is an indication of the length of time life forms have lived in that isolated situation. This principle is used by biologists and paleontologists as a clue to the past history of various environments. Australia, for example, must have been isolated for many millions of years (perhaps thirty-five to fifty million), judging from the unusual collection of animals discovered when ships bearing Europeans first landed on that continent. Almost all the niches occupied by placental mammals in other parts of the world were occupied by marsupial mammals and egg-laying mono-tremes: the koala bear, the bandicoot, the kangaroo, the wombat, the duck-billed platypus, and many more equally outlandish creatures.

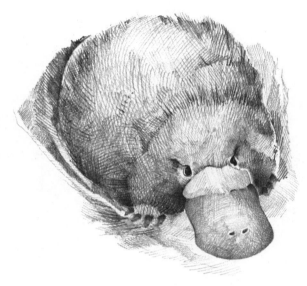

South America still harbors some endemic species today, but the period of great isolation seems to have occurred earlier in its history. Fossils between sixty-five and three million years old provide proof that a highly distinctive land fauna evolved there: giant sloths, armadillos, anteaters, and many marsupial mammals similar to those found later in Australia but not identical with them. These animals roamed the South American continent undisturbed by outside influences for many eons. Then, about three million years ago volcanic activity created a land connection between North and South America and northern fauna invaded the southern continent. Many of the endemic species there became extinct.

Biological facts like these are used in reconstructing the arrangement of land and sea at various periods in the planet's history. The presence of identical species and communities of terrestrial organisms suggests contiguous landmasses; development of endemic and highly specialized organisms suggests isolation of that region. The principles are simple enough, but there are many fine points that must be considered in applying them. For example, we cannot assume that isolation of land animals always requires the existence of an ocean barrier. They may also be isolated by high mountain ranges, deserts, or great differences in climate.

Aquatic organisms are isolated by landmasses, but deep ocean is an equally effective barrier for shallow-water species. Climate and food supplies limit the range of organisms in the sea as they do on land although the climatic variations are less extreme in the ocean. Finally, it must be remembered that some dispersion occurs by wind and water. If lands are nearby but not contiguous, the small, mobile species may bridge the gap by hitching a ride on floating rafts of vegetation. Birds, flies, and even spiders can ride on the wind.

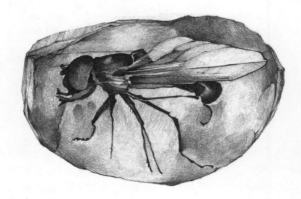

Using the information from fossil remains, from paleomagnetic measurements, from dating of rocks, from matching rock formations, and from evidence of similar climatic zones, earth scientists have been attempting to fit back together again the changing shapes of the landmasses throughout geologic time. There is general agreement, except for a few small details, in the reconstruction for 200 million years ago. It is strikingly similar to the map that Wegener drew in 1912 — a world consisting of one large continent, Pangaea, and one great world ocean, the proto-Pacific.

But before Pangaea what? Did plate tectonics begin with the break-up of that single continent? Most geologists believe not. They think that crustal movements may have begun as soon as a hard surface formed on the planet — at least three and a half billion years ago. But as the search goes further and further back in earth history, the evidence defining certain land and sea configurations becomes more tenuous. Sometimes biological evidence seems to conflict with paleo-magnetic evidence. Climate information must be reconciled with relationships of land and sea as well as with latitude. A solution accommodating all types of data must be sought, presenting a tremendous and intriguing challenge to the imagination.

Alfred Ziegler, Christopher Scotese, and several colleagues at the University of Chicago have produced the most complete and detailed reconstructions of the way the earth looked in ages past. These maps are the result of many years of work. Even now they are not considered finished but, like Wegener's first reconstruction, are working models that will be modified as further information becomes available.

Six of these maps are reproduced on the following pages, beginning with the recent past and working back into more and more ancient history (Figure 3). As you can see, the familiar shapes becomes less recognizable as the years spin backward. Sixty million years B.P. the southern Atlantic was much narrower than it is today; Italy, Greece, and Yugoslavia constituted a small land mass disconnected from Europe. India was floating free thousands of miles south of its present position. Australia, New Zealand, and Antarctica were joined together. Two hundred million B.P. the landmasses were all united in a single continent — Pangaea. Three hundred million B.P. there were many smaller landmasses. China was quite separate from Siberia and was stretched out into a long, tenuously connected piece. Parts of Russia, Turkey, and the Middle East made up a separate continent, named Kazakhstania by the authors of these maps. Five hundred million B.P. the landmasses were even more widely distributed. Europe lay on the frigid Antarctic circle while Siberia basked in tropical sunshine. South America was upside down with Tierra del Fuego intercepting the equator. What a lot of "impossible things"! The White Queen would have been delighted with these maps.

In order to handle the complex data used in the reconstructions a number of simplifying assumptions had to be made and it is important to keep these in mind. It was assumed that the magnetic poles have remained very close to the present poles of rotation throughout earth history. North and south occasionally exchange positions, as we have seen, but the magnetic axis is assumed to be more or less coincident with the axis of rotation. Actually, we know that this is not strictly true. The present south magnetic pole is approximately 24 degrees away from true south. Furthermore, the position of the magnetic poles has varied within recorded history. The changes have been relatively small, however, and they seem to have averaged out over time. Magnetic

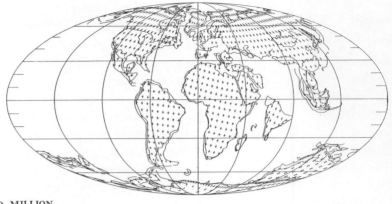

60 MILLION

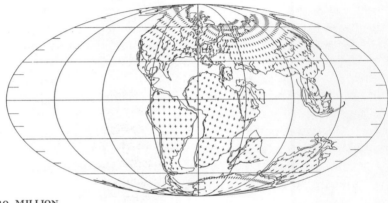

120 MILLION

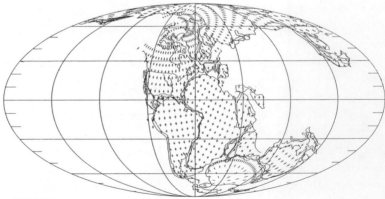

200 MILLION

FIGURE 3 *Paleographic maps. Mollweide projection, front view is used.*
Small crosses show present orientation of latitude and longitude, in-

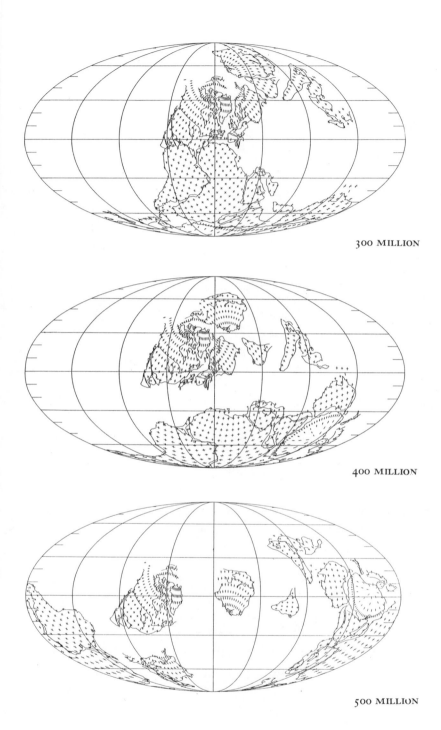

300 MILLION

400 MILLION

500 MILLION

dicating rotation of the land mass throughout time. (Copyright by A. M. Ziegler and C. R. Scotese, reprinted by permission) data would be very ambiguous and almost impossible to interpret without the simplifying assumption that the magnetic axis and the axis of rotation have remained approximately coincident.

The second important assumption is that the radius of the planet has remained constant. This is one of the cornerstones of plate tectonics and is generally accepted by the vast majority of earth scientists. In the absence of any definite evidence of contraction or expansion they believe that the constant radius is the simplest and therefore the best hypothesis. But there are minority opinions on this subject, and the history of science has taught us to respect minority points of view.

The hypothesis of a shrinking earth was actually the majority opinion just a few decades ago. It provided an explanation for the formation of mountain ranges: as the planet shrank because of gradual cooling of the original hot ball, the crust crumpled and wrinkled like the skin of a prune. Plate tectonics has a different explanation of mountain ranges: two plates in collision cause the lithosphere to buckle. So there is no longer any need to postulate shrinkage of the earth. Nevertheless, there are some geophysicists who still believe that some reduction in earth radius is occurring even at the present time.

There are also a few earth scientists who think that the recent discoveries of seafloor spreading and the formation of new crust can best be explained as the result of an expansion process. S. Warren Carey, one of the principal advocates of this theory, points out that expansion is known to be occurring in many places. Subduction, on the other hand, is a hypothesis based on theoretical considerations and has not yet been proved. The rate of expansion measured in spreading rifts around the world would have been sufficient to create all the seafloor basins in 200 million years (the period of time within which they are all dated). Pangaea, Carey suggests, might have been the entire crust of the earth at that time and the radius of the planet would have been about two thirds the present radius.

This idea, which is an anathema to most earth scientists, might receive a more sympathetic hearing if its advocates could suggest a plausible physical mechanism for such an expansion. (Remember, it was the lack of physical explanation that blocked the acceptance of Wegener's idea, too.) Although several theories for increase in radius have been proposed — decreasing density, a very small but steady decrease in the gravitational constant — none of these theories envisions an expansion of anything like the magnitude proposed by Carey.

Fortunately there are today sophisticated techniques capable of yielding measurements that should soon be able to settle this question as well as provide direct measurement of the rates at which various portions of the earth's crust are moving around the planet. Triangulation techniques similar to those that Wegener used on the ill-fated Greenland expedition have now been perfected to the point where they can detect very small changes in distances on the earth's surface — for example, an increase or decrease of a few inches in the space between New York and Paris. Such information averaged over a number of years can tell us whether there are regular changes in these distances. The methods utilize signals from satellites and from astronomical sources as Wegener attempted. A signal from a single source arrives at two stations on earth at slightly different times. The time differences can be very accurately recorded by atomic clocks and translated into a measurement of the ground distance between the two receiving stations.

At the Jet Propulsion Laboratory in Pasadena and at Goddard Space Flight Center in Maryland groups of geodesists are working on a method using radio waves from quasars — those amazingly powerful stellarlike objects that are traveling at great speeds away from us toward the edge of the universe. Thus radiation that has taken billions of years to reach us may tell us what is happening beneath our very feet.

Another method is based on the use of satellites. Pulses of sharply focused laser light are beamed from ground stations to a satellite passing overhead. The signal is reflected back to the station and the round-trip time is used to compute distance.

In a few more years the research teams working on these two projects expect to have achieved the necessary degree of accuracy and to have collected data over a long enough period of time to produce significant information on movements of the earth's crust, especially in those places where the changes are occurring most rapidly. In the satellite project measurements are now being collected from sixteen receiving stations in key locations around the world: Samoa, the Marshall Islands, Kauai, Australia, Peru, Japan, and a number of places in the United States and Europe. The scientists hope to have access soon to stations in Easter Island, Mexico, and Tahiti.

Enigmatic little Easter Island may be the first place on earth to confirm or refute the estimates of plate tectonic theory about crustal movement. It lies on the East Pacific Rise, the fastest spreading piece of seafloor yet discovered. According to theory, Easter Island should be moving toward South America at the average rate of 6.3 inches a year and a change of this magnitude is well within the limits of accuracy that the project has already achieved.

Wegener would have been happy indeed if he could have taken part in these experiments — to measure the girth of the planet and to calculate the restless wandering of continents around the globe, to see the creations of his imagination transformed into fact, perhaps also to witness the birth of new mysteries as the old ones are laid to rest. When the results are all in, the White Queen's radical ideas may be vindicated, too. We may have to believe several impossible things before breakfast.

Fires of Creation

In all ages volcanoes have frightened,
fascinated and attracted man, because what
they hold is at once terrifying, splendid, and
mysterious.

— HAROUN TAZIEFF,
Craters of Fire

The Greek philosopher Empedocles, who was born five centuries before Christ in Sicily, less than a hundred miles from the largest volcano in Europe, was one of the first men to suggest that nature can be understood. The basis for understanding it, he believed, was the discovery of an order that lies beneath the bewildering variety that we observe in the external world. All things, Empedocles said, can be reduced to four primary substances: fire, earth, air, and water. It is the coming together and parting of these elements that make up the great diversity of the physical world. In an age when many thinkers were rejecting the role of observation and the senses, Empedocles taught that the senses are avenues to understanding. Working from these simple premises, he made several surprisingly accurate deductions: light needs time to reach us from the sun; night is caused by the earth's intercepting the rays of sunlight; plants and animals have evolved into higher and more complex forms by a selective process. Those types that have best met the conditions of life have survived in the greatest numbers. Thus Empedocles can rank as one of the world's earliest scientists.

Empedocles may have been influenced in his choice of primal substances by his close association with and long-time observation of the great volcano of Mount Etna, which dominated his native land. According to one story, he met his end by tumbling into the flaming crater. Perhaps he had ventured too close in an effort to observe more accurately the explosive coming together and violent parting of earth, air, fire, and water. We can imagine that he was striving to understand the power that molds mountain ranges, carves out new river channels, spins the gossamer fabric of clouds from air and water, and piles up the thunderheads where fire springs forth again and plunges back toward earth.

We will never know what new insights may have come to Empedocles in those last hours, face to face with this mighty and humbling power. The volcano swallowed him up, the legend says; all that was left of him was his sandals.

Most people today would agree with Empedocles that nature can be understood, at least in its quiet, day-to-day activities: the progress of the sun across the sky each hour, the regular turning of the seasons, the rhythmic rise and fall of the tides, the paths that the planets trace throughout the year. But violent and destructive phenomena, like a volcanic eruption or an earthquake or an avalanche, are characterized as "nature gone wild." Our response to these events is more like that of primitive people invoking supernatural causes, like the Northwest Indian tribes who believed that volcanoes were caused by giants locked in angry combat; or the Polynesians in Hawaii whose volcanic deity, Pele, must be constantly propitiated; or the Irish monks of the twelfth century who described the volcano of Heklafell on Iceland as the "Gate of Hell," a place of everlasting fire where souls suffer for eternity. "Whenever great battles are fought," they said, "or there is bloody carnage somewhere on the globe, then there may be heard in the mountain fearful howlings, weeping, and gnashing of teeth."

When Mount St. Helens erupted in May 1980, local inhabitants were outraged. ("Our beautiful mountain has betrayed us!") And they were afraid. ("You felt like falling down on your knees and covering your face!")

For a moment, let us put on the empty sandals of Empedocles, look down into the roaring craters where new earth, new water, and new air are being created in a fiery furnace. Let us consider whether this great process can be understood, measured, even predicted. Is it a strange, outrageous aberration, an abnormal condition of nature? Or is it, as Empedocles probably thought, a dramatic manifestation of the primal forces that have shaped and are still shaping our world?

In the last few centuries volcanic eruptions of many types have been studied and monitored, sometimes at considerable risk to the observers. These eruptions have run the gamut from the gentle Hawaiian type to devastating fire-clouds of destruction. They have involved many established volcanoes like Mount St. Helens, but brand-new volcanoes have also been born in historical times.

~❧

On February 20, 1943, Dionisio Pulido walked with his wife Paula, his son, and a neighbor from the village of Paricutín to the small plot of land he owned in the valley of Rancho Tepacua. It was time to prepare the land for the spring planting of corn. Pulido, his family, and friend worked all day in the blazing Mexican sun. In the late afternoon when the mountain peaks to the west began to cast long shadows across the little nine-acre farm, Pulido looked back across the field and noticed a crack in the ground about half a yard deep. It spoiled the uniform contours of his newly plowed field. He pushed a few stones in to fill it and smoothed out the dirt. But he had no sooner turned his back and walked a few paces than there was a deafening roar like thunder. The ground shook, the trees trembled, and when he looked back he saw that where the crack had been, the earth had risen up into a giant blister higher than a man's head. As Pulido and his family watched, smoke and fine gray dust began to rise in a plume that grew ever higher and darker. A loud and continuous hissing sound and a strong, choking odor of sulfur filled the air. Sparks began to pour out of the fissure and pine trees one hundred feet away suddenly burst into flame. The four eyewitnesses to the birth of a new volcano fled from the scene and ran the two miles back to their village.

The next morning Dionisio Pulido returned to his fields and found a cone almost thirty feet high spitting forth ash and rocks with great violence. By midday the cone had grown to a hundred feet in height. As the days passed, the cone continued to pile up at an amazing rate. Pulido's small acreage was completely buried under a mass of black slag. In the evenings, watching from the village of Paricutín, Pulido and his neighbors saw tons of incandescent rocks thrown thousands of feet into the air. Like giant fireworks, the rocks reached the end of their trajectory, seemed to pause for a moment, then fell back and shattered on the cone, cascading down the slope in interlacing rivers of fire. Soon the village had to be evacuated before it, too, was buried in lava. In the larger town of San Juan de Parangaricutiro, which lay three miles from the volcano, nothing was spared but the church steeple.

The eruption of Paricutín did not claim any lives. It occurred in a sparsely inhabited area and was a type of eruption that proceeds in a relatively predictable manner. With a little prudence the local inhabitants were able to maintain a safe distance.

There are localities in the world where volcanoes are even less threatening than Paricutín. The Hawaiian Islands owe their whole existence to volcanic activity. They boiled up out of the sea several million years ago. The oldest island lies farthest to the northwest. The others were formed successively in a long chain extending southeast to the newest island, Hawaii, where the volcanoes Mauna Loa and Kilauea sometimes spurt forth fountains of fire. Even during these active periods, observers can usually come right up to the lip of the crater without danger, look down into lakes of glowing lava, and watch long ribbons of fire descending the flanks of the mountain.

These Hawaiian volcanoes are relatively benign because the basaltic lava that erupts from them is very fluid. The violence of a volcanic eruption is determined in part by the chemistry of the magma and the amount of gas that is dissolved in this semimolten matter under high-pressure conditions below the surface of the earth. As the hot magma rises and the pressure decreases, the gas dissolved in it comes out of solution and tries to escape. If the magma is very fluid, the gas can escape easily and does not collect in pockets. Viscous magma, on the other hand, tends to trap the gas, which then collects until there is enough pressure to fracture the overlying rock and a violent explosion takes place. The nature of the volcanic eruption depends upon the composition of the particular magma, the amount of gas present, and the shape of the fissures. Several grades of explosiveness have been classified and named after famous eruptions.

The Strombolian type involves basaltic lava like the Hawaiian eruptions, but it is a little more explosive. The volcano regularly spits out hot fragments of semisolid material that fall back to earth, accumulating in a scoria cone, which usually has a symmetrical shape with sides pitched at 33 degrees, the angle of rest for loose stones. Paricutín was a Strombolian eruption.

Vulcanian eruptions eject more viscous lava and tend to be more violent than the Strombolian. Parts of the volcanic cone are demolished. Large solid pieces and highly fragmented ash are thrown to great heights in intermittent explosions that produce tall plumes of gas and dark clouds of volcanic ash.

Vesuvian eruptions are similar to Vulcanian, but they release more energy. The activity is continuous, carrying ash even higher into the atmosphere. The ash is fragmented magma rather than shattered bits of old rock. Eruptions of this type, especially those that create unusually large amounts of ash, are sometimes called Plinian, in memory of the extraordinary firsthand account written by Pliny the Younger, describing the famous eruption of Vesuvius in A.D. 79 and the death of his uncle, Pliny the Elder:

. . . My uncle was stationed at Misenum [about twenty miles from Vesuvius] in active command of the fleet. On 24 August, in the early afternoon, my mother drew his attention to a cloud of unusual size and appearance. He had been out in the sun, had taken a cold bath, and lunched while lying down, and was then working at his books. He called for his shoes and climbed up to a place which would give him the best view of the phenomenon. It was not clear at that distance from which mountain the cloud was rising (it was afterwards known to be Vesuvius); its general appearance can best be expressed as being like an umbrella pine, for it rose to a great height on a sort of trunk and then split off into branches. . . . Sometimes it looked white, sometimes blotched and dirty, according to the amount of soil and ashes it carried with it. My uncle's scholarly acumen saw at once that it was important enough for a closer inspection, and he ordered a fast boat to be made ready, telling me I could come with him if I wished. I replied that I preferred to go on with my studies, and as it happened he had himself given me some writing to do.

As he was leaving the house he was handed a message from Rectina, wife of Tascius whose house was at the foot of the mountain, so that escape was impossible except by boat. She was terrified by the danger threatening her and implored him to rescue her from her fate. He changed his plans, and what he had begun in a spirit of inquiry he completed as a hero. He gave orders for the warships to be launched and went on board himself with the intention of bringing help to many more people besides Rectina, for this lovely stretch of coast was thickly populated. He hurried to the place which everyone else was hastily leaving, steering his course straight for the danger zone. He was entirely fearless, describing each new movement and phase of the portent to be noted down exactly as he observed them. Ashes were already falling, hotter and thicker as the ships drew near, followed by bits of pumice and blackened stones, charred and cracked by the flames; then suddenly they were in shallow water, and the shore was blocked by the debris from the mountain. For a moment my uncle wondered whether to turn back, but when the helmsman advised this he refused, telling him that Fortune stood by the courageous and they must make for Pomponianus [a friend] at Stabiae. . . . Pomponianus had . . . already put his belongings on board ship, intending to escape if the contrary wind fell. This wind was of course full in my uncle's favour, and he was able to bring his ship in. . . .

Meanwhile on Mount Vesuvius broad sheets of fire and leaping flames blazed at several points, their bright glare emphasized by the darkness of night. My uncle tried to allay the fears of his companions by repeatedly declaring that these were nothing but bonfires left by the peasants in their terror, or else empty houses on fire in the districts they had abandoned. Then he went to rest and certainly slept, for as he was a stout man his breathing was rather loud and heavy and could be heard by people coming and going outside his door. By this time the courtyard giving access to his room was full of ashes mixed with pumice-stones, so that its level had risen, and if he had stayed in the room any longer he would never have got out. He was wakened, came out and joined Pomponianus and the rest of the household who had sat up all night. They debated whether to stay indoors or take their chance in the open, for the buildings were now shaking with violent shocks, and seemed to be swaying to and fro as if they were torn from their foundations. Outside on the other hand, there was the danger of falling pumice-stones, even though these were light and porous; however, after comparing the risks they chose the latter. . . . As a protection against falling objects they put pillows on their heads tied down with cloths.

Elsewhere there was daylight by this time, but they were still in darkness, blacker and denser than any night that ever was. . . . My uncle decided to go down to the shore and investigate on the spot the possibility of any escape by sea, but he found the waves still wild and dangerous. A sheet was spread on the ground for him to lie down, and he repeatedly asked for cold water to drink. Then the flames and smell of sulphur which gave warning of the approaching fire drove the others to take flight and roused him to stand up. He stood leaning on two slaves and then suddenly collapsed, I imagine because the dense fumes choked his breathing by blocking his windpipe which was constitutionally weak and narrow and often inflamed. When daylight returned on the 26th — two days after the last day he had seen — his body was found intact and uninjured, still fully clothed and looking more like sleep than death. . . .

This description written by a seventeen-year-old Roman boy is the earliest known eyewitness account of a volcanic eruption and can be regarded as the beginning of the science of volcanology.

&

The eruption in 1883 of Krakatoa between the Indonesian islands of Java and Sumatra was one of the most destructive volcanic explosions of recent times. It has given its name to the catastrophic events that occur when major portions of the mountain collapse, creating an enormous caldera. If they occur in the ocean, they set up waves called tsunamis that carry the shock and devastation to far distant shores.

A Peléean eruption, however, is even more fearful than a Krakatoan because it produces an incandescent cloud of rock fragments and very hot gases that is blasted sideways out of the vent and rolls down the mountain at terrifying speeds. The solid material is cushioned and borne along by the exploding gas. This *nuée ardente*, as it is called, burns and destroys almost everything in its path like a fire cloud of a nuclear bomb.

Mount Pelée erupted on May 8, 1902. Before that terrible day the beautiful, symmetrical cone of the mountain had stood 13,000 feet above the port city of St. Pierre, Martinique, in the Windward Islands. The mountain occasionally trailed little wisps of steam but not enough to cause any alarm. Local inhabitants hiked and picnicked on its cool slopes.

Around the middle of April 1902 the mountain showed signs of awakening. Bigger and darker plumes poured out with sulfurous odor, accompanied by ominous rumblings and small earthquakes. Snakes that normally inhabited the old lava fields near the summit left the mountaintop and began invading the plantations in the valleys. Flocks of migrating birds that often flew over the mountain were observed to avoid the area. Sailors and fishermen noticed regions in the ocean where warm water was welling up and ground swells occurred in clear calm weather. Late in April fine ash began to sift down on the countryside. On May 5, a torrent of steaming mud poured down the mountain, killing twenty-four people and warning of an eruption in progress. But the citizens of St. Pierre were more interested in politics that week. An election day was coming up on May 10 and the authorities were anxious to keep the population together until after the election was over. Reassuring announcements about the volcano were posted. A scientific commission was appointed and it produced the desired report, setting to rest the fears of the populace. Only a relatively few people left St. Pierre and many more sought refuge there from the surrounding countryside. The population was swollen to nearly 30,000.

During May 6 and 7 dense clouds of smoke erupted from the mountain; black curtains of ash descended over the town and far out at sea. The noise of the eruptions was terrifying. May 8, Ascension Day, was a day without a dawn. In the early morning hours panic broke out in the town. Thousands of men, women, and children left their houses and crowded onto the waterfront, instinctively seeking safety by the sea. Flames were now leaping from the crater and suddenly a deafening series of detonations split the air like the thunder of a thousand cannons. The top of the mountain burst open. A great dark cloud emerged and rolled swiftly down the hill. It was thick and purple-colored, lit from within with lightning flashes. In just three seconds the cloud covered the distance from the mountaintop to the outskirts of St. Pierre. In the next second the whole town disappeared in the burning cloud. The pressure of the advancing air drove the entire mass of frightened people huddled on the waterfront into the sea. The water of the harbor boiled, clouds of steam rose, and ships were capsized or turned into flaming torches. The climax of the conflagration was reached when thousands of casks of rum stored in warehouses exploded and burned. The fiery brew flowed in rivers through streets where houses were also ablaze. Some of it found its way into the sea and spread a burning film out across the waters, setting many more ships afire.

The sheer force of the volcanic cloud that struck St. Pierre knocked down walls three feet thick, tore heavy cannons from their mounts. Masses of roofing were ripped loose from the houses and were found wrapped like cloth around posts and trees against which they had been flung. The heat of the cloud was so intense that many bodies were badly burned though their clothes were not even singed.

Out of 30,000 only two people survived the tragedy. One lived at the extreme outskirts of the region seared by the *nuée ardente*. The other was locked up in a windowless cell underground in the town jail. Severely burned even so, this convicted murderer lived through four days without food before he was rescued and survived to tell the story of the Peléean eruption that wiped out St. Pierre.

⤫

But the most dreadful type of volcanic eruption has not yet been named. Although the evidence for these fantastically powerful eruptions is widespread, an event of this kind has never been directly observed during recorded human history. Rocks called ignimbrite, the product of these eruptions, occur over many thousands of square miles in Nevada, England, Italy, South America, Mexico, and New Zealand — to cite just a few. Ignimbrite is frothy rock made of volcanic glass, similar to pumice. It also contains many pieces of solid lava, loose crystals, and irregular fragments, welded together by heat and the pressure of overlying layers of rock. The distinctive texture of ignimbrite can easily be recognized. Because of its lightness and great strength it is valued as a building stone.

Eruptions that produce ignimbrites are thought to occur in situations where the magma is very viscous but degassing occurs near the surface so the hot, expanding mass foams out horizontally like the *nuée ardente* and travels down the mountainside at incredible speeds. But while the *nuées ardentes* are mostly gas and usually do not reach farther than the base of the mountain, ignimbrites carry enormous quantities of solid material to great distances around the volcano.

One ignimbrite eruption has occurred in recent times; however, it took place in a remote, uninhabited part of Alaska. No one actually observed the blowing up of Mount Katmai in June 1912, although at settlements about 100 miles away day was turned into night for almost a week. Several years later a scientific team explored the area. They found a great valley filled to an average depth of 100 feet with finely fragmented ignimbrite, about the size of coarse rice. Over an area of thirty square miles the ground was all broken open and innumerable plumes of hot gases were pouring forth from the fissures. Robert Griggs, leader of the expedition, said: "It was as though all the steam engines of the world, assembled together, had popped their safety valves at once, and were letting off steam in concert." He named this place the "Valley of Ten Thousand Smokes."

Griggs estimated that if such an eruption had occurred in downtown New York City, the whole of Manhattan Island and an equal area besides would have been swallowed up in a field of smoking chasms. Red-hot sand would have run like wildfire through the rest of the city, burning or vaporizing everything it touched. Many of the tallest buildings would be completely buried in the flow of incandescent rock. A hole would be blown in the ground big enough to contain all the buildings of greater New York several times over. If the winds were right, Philadelphia would be blanketed with ash a foot deep and even as far away as Toronto acid rain would cause burns on exposed parts of the body.

Such a projection sounds like a description of nuclear holocaust, not an act of nature. But the Katmai explosion was a small ignimbrite eruption compared with some that have been identified in the geologic record. The eruption that produced Crater Lake in Oregon was an ignimbrite event that spewed out somewhere between three and four cubic miles of rock and ash. An eruption that occurred near Naples approximately a million years ago covered 1,200 square miles and the ignimbrite flowed uphill into mountain passes as high as 2,000 feet.

A chilling thought is suggested by these facts. Will ignimbrite eruptions be called someday by the name of a mountain that is familiar to us all? Mount Hood? El Fuego? Popocatepetl? Or will they bear a name unknown to us now — an obscure mountain village, a field where a farmer plows his crops within sight of the sea, never dreaming of the vast body of hot rock that breathes and stirs just a few miles beneath his feet?

❧

The knowledge that has been gathered about volcanoes suggests that eruptions are most likely to occur in or near the ocean. Almost all volcanoes are either in the sea or no farther than 150 miles from its edge. The few notable exceptions are in the African Rift Valley. In general, volcanic action is concentrated along the boundaries of lithospheric plates (see Figures 2 and 4). It occurs at midocean ridges where plates diverge and seafloor spreading is taking place. It also occurs along the boundaries where plates are colliding and some crustal material is being forced down again into the mantle. As this material descends to great depths, it is heated and becomes partially melted. Wherever this semiliquid magma finds a crack or weak spot in the crust it moves upward again. Hot gases are released and a new fissure may be opened up. But as the magma moves through large thicknesses of continental crust to reach the surface, some of the rock it meets is also melted and incorporated into the hot material. These additions change the chemical composition of the magma, making it more viscous and increasing the violence of the explosion. Therefore dangerous eruptions are most apt to occur on land near the meeting place of an oceanic and continental plate. The borders of the Pacific are believed to coincide with the boundaries of converging plates, thus creating the Ring of Fire.

The ocean floor, as we have already seen, contains an almost incredible number of volcanoes. The total amount of lava extruded from volcanoes in the Pacific basin alone during the last 100 million years is much greater than that erupted by volcanoes on all the continents during the same period. It is estimated that at least 10,000 volcanoes with a height of half a mile or more exist in the Pacific. Some reach high enough to emerge from the sea and continue to pile up into the steep cones of volcanic islands. In the Hawaiian Islands, lava has created mountains that soar 31,000 feet from the ocean floor, making these the tallest mounds of matter on earth.

The sudden appearance, erosion by wave action, and subsequent drowning of islands has actually been observed several times in recent centuries. One of the classic examples was an island that erupted from the sea southwest of Sicily and then disappeared again a few months later. On June 28, 1831, the crew of a British ship in that area reported that they felt a sudden shock as though the ship had run aground. But subsequent observations showed that they had experienced an earthquake at sea. About two weeks later the captain of a Sicilian vessel en route to Agrigento (birthplace of Empedocles) in Sicily saw a tremendous column of steam rising from the ocean like an enormous waterspout. On July 18, returning from Agrigento, the sailors passed

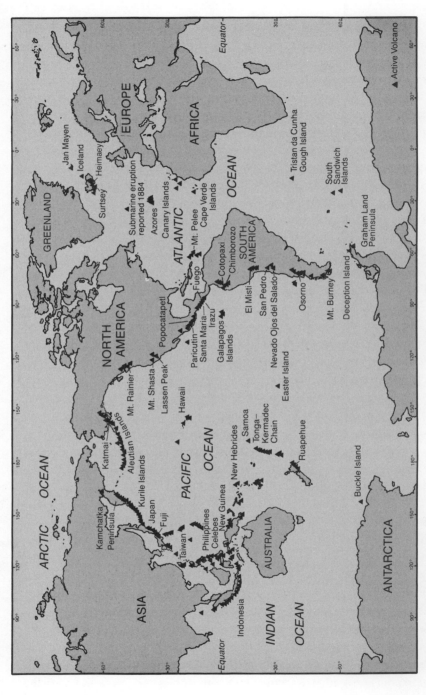

FIGURE 4 *The distribution of active volcanoes. (Redrawn by George Ward, from Peter Francis, Volcanoes, Pelican Books, 1976, Fig. 1 [Parts 1 and 2]: 16–17. Copyright © Peter Francis, 1976. Used by permission of Penguin Books Ltd.)*

the same spot and found that an island had risen out of the sea. It was about twelve feet high with a volcanic crater in the center ejecting great plumes of vapor and projectiles of molten rock. Inside the circular crater a pool of murky red liquid boiled and seethed. By August 4 the island was 300 feet high and several miles in circumference. After that time the volcanic activity of the main vent subsided, but violent agitation of the sea southwest of the island occurred with columns of dense white steam. Geologists concluded that a second vent had erupted but had not ejected enough solid matter to reach the surface. In the meantime the waves, the wind, and the rain beating on the small new island were rapidly eroding its surface and carrying it back to the sea. By the end of October there was only a little mound of volcanic rock left above water, and two years later it was entirely submerged.

In recent years many drowned islands like this have been observed below the surface of the ocean. In 1945 at the height of World War II a U.S. troop transport was crossing the cold, bleak waters near the Aleutian Islands. With its fathometer it was constantly monitoring the depth of the ocean bottom. The transport's commander, Harry H. Hess, observed on several occasions some strange signals on the fathometer scope. The reflections seemed to come from large flat-topped cones rising up from the ocean floor. Hess was a geologist taking time off from his profession to fight the war and he recognized that these objects must be submerged mountains.

After the war he followed up this discovery. Eventually 160 very large truncated mountains were identified in the Pacific. Their tops ranged from 3,000 to 6,000 feet below present sea level. Each of them had a flattened summit as though the peak of a cone had been sliced off. Hess suggested that these seamounts (or guyots, as he called them) were old volcanic islands. When sea level had reached within a few hundred yards of the peak, wave action began to erode the land that was still emergent and gradually the peak was planed off. Continued subsidence drowned the islands completely.

⤳

Igneous activity has also shaped and molded the continents of the earth. Peel back the layers of rock almost anywhere on the land and you will find evidence of the rifting, melting, and decrystallizing that have taken place as the thin crust of the planet has moved in response to currents in the warm, semiliquid material of the upper mantle. The crust itself originally solidified from magma. Although strictly speaking, "volcanism" refers to the emission of molten lava and gases through vents in the planet's crust, this activity is part of a more general process by which liquid magma becomes solidified at or below the earth's surface, creating the entire profile and complexion of the planet.

An active volcano spreads new earth on its mountainsides and scatters it broadcast in fine grains over the surrounding valleys. This earth is rich in many essential ingredients: lime, potash, and phosphates. These ingredients are slowly leached out of older soils and must be replaced to retain fertility. As volcanic ash weathers, it makes a rich, friable soil in which many types of vegetation flourish. The best coffee and tea in the world are grown on the slopes of ancient volcanoes — the coffee plantations of Guatemala and Costa Rica, the tea plantations of Sri Lanka. In the protected valleys at the base of the mountains, ash deposits have periodically renewed the earth so warmth-loving crops like rice and coconut palms grow well there. The slopes at the base of Mount Etna are luxuriantly fertile, producing fine crops of grapes and olives and other fruit. Vegetables and corn often yield five harvests a year. One cannot visit regions of recent volcanic action like Sicily and Kauai without being deeply impressed by the richness of the vegetation. Brilliantly colored flowers and fruits grow in exuberant profusion.

Within a few decades after a major eruption the typical shield volcano is transformed like Cinderella from a dirty ash heap into one of the most beautiful and desirable landscape features on earth. Its lush lower slopes give way to deeply carpeted mountain meadows and finally to a glistening crown of year-round snow. The mountain's varied climate zones provide a spectrum of splendid places to live — places with cool, clean air, abundant crops, sunshine, and panoramic views across the countryside. So the quiescent volcano is a baited trap luring people back to its slopes long before it is safe to assume that the volcano is truly dormant. Shortly after the last ashes have fallen, people start coming back and erecting their frail homes again on these treacherous slopes.

In 1970 I saw new houses being built on lava fields near the summit of Mount Etna. At that time the volcano was frequently shooting off little fountains of fire from her snow-covered peak. From the town of Taormina I could watch the pretty displays almost every night. Just a year later a major eruption of Mount Etna wiped out the new houses that had been built with such naive optimism.

Many similar examples can be found throughout the world. In the Philippine Islands about thirty miles south of Manila, there is a lake with a tiny island named Taal. This island contains the crater of one of the most dangerous volcanoes in the world. It has erupted frequently in the past 400 years, claiming many lives. In 1911 an eruption killed all but 15 of the 500 people living on the island and 800 more on the shores of the lake. Appalling scenes of violent death and disaster were still vivid in the minds of the local people but, like moths to a flame, they were attracted back to the mouth of the inferno. Soon the very flanks of the volcano were repopulated. In 1965 another violent eruption occurred. In spite of the fact that a warning had been issued when the temperature of the lake rose above normal values, 150 people were killed. A few weeks after the eruption was over, people began returning to the island. There are now more people on Taal than there were at the turn of the century.

The reoccupation of the land usually begins after the ash flow is no longer a threat to agriculture. At first the ash is damaging to crops because it blankets the leaves, deters photosynthesis, and smothers the plants. In some cases it contains significant amounts of poisonous elements, such as fluorine. The length of time required for the ash to weather and become incorporated into the soil is determined by the depth of the ash flow and the amount of rainfall. With a thin ash deposit and abundant moisture the soil can recover in a few weeks or months. Where the ash is many inches deep, particularly in dry regions, the soil may not be usable for a generation. But eventually the land recovers and becomes more productive than it was before the eruption.

Volcanic action also serves to isolate and concentrate some of the most valuable minerals within the earth's crust — gold, silver, and gemstones like rubies and diamonds — by a wonderful and somewhat mysterious alchemy that we will examine in a later chapter.

It has created the soft, protective layer of atmosphere that envelops the earth. The clouds of steam and gases ejected from the volcanic craters contain nitrogen, carbon dioxide, sulfur dioxide, hydrogen sulfide, water vapor, and some trace elements. Many of the gases have been distilled from the hot magma and released from the deep interior of the earth. Gradually over the billions of years since the planet first condensed from the cosmic dust cloud these breaths of new air have reached the surface and joined the other gases of the atmosphere. Reacting with one another, with the rocky crust and living things, they have slowly evolved into the complex, sensitive medium that now wraps our planet in a delicate veil of blue.

The clouds, too, and the water of the seas have come from the warm inner layers of the earth. The plumes of steam that erupt from volcanoes contain a small amount of "juvenile" water, which has never been on the earth's surface before, as well as recirculated water that had been on the surface at one time but has found its way down again through the crust. Although the exact proportions are still being debated, on one point there is universal agreement: if there had been no release of volatile elements from the solid body of the planet, the earth would have been uninhabitable like Mars or the moon. Life as we know it could not have evolved on a waterless planet.

Indeed, according to one theory, life itself may have been conceived in a volcanic eruption. In and around the dark cumulus clouds that accompany eruptions, intense electrical storms rage. The air for miles around is electrified. During the May 18, 1980, eruption of Mount St. Helens sparks jumped from ice picks of mountain climbers thirty miles away. Perhaps, so the theory goes, lightning and the heat associated with volcanic eruptions provided the energy that synthesized the complex protein particles which serve as the precursors of the simplest living things.

Most of these facts were unknown to Empedocles as he stood looking down into the crater of Mount Etna. In its ribbons of incandescent matter, its fierce heat, and its hissing voice he saw revealed something tremendously dynamic, glowing with naked power. Yet he stood fast in his belief that nature could be understood. The volcano is not a dragon breathing fire nor the gate to Hell. It is a tiny window into the forge where the planet we live upon is being constantly reshaped and made anew. The process is awesome; its immensity dwarfs our human scale, but still it can be understood. As we watch the meeting and parting of primal elements we, too, may discover order even in extreme violence and apparent chaos. We may catch intimations of the unifying principles underlying all nature and predict our planet's future as it spins on its intricate way through space and time.

6

The Planet Breathes and Stirs

I am he who moves the earth
[and] will demolish all the
world.

— *Popul Vuh*, Mayan Scripture

In the early weeks of December 1974 the Chinese city of Haicheng lay hushed and muted beneath the brittle crust of winter. Icy winds sweeping down from Inner Mongolia had turned the soft earth of the Liaotung peninsula to a hard frozen shell dusted with snow and had spread thick sheets of ice on all the little ponds and lakes around the city. By day the people of this busy industrial city moved quickly and quietly through the streets as they went to and from work. By night they retreated into their brick and adobe buildings, shutting the doors against the cold blasts. Often only the sound of lorries rattling over the rough streets broke the winter stillness.

But toward the end of December a thread of new sounds and an undercurrent of nervous activity began to disturb the quiet scene. Pet cats — normally gentle and sleepy — howled dismally in the night and refused to be quieted. Dogs wandered through the city barking wildly. Rats that usually stayed hidden during daylight hours were seen scurrying boldly in packs through the streets. Domestic ducks and chickens fluttered about, cackling and honking as though they were being chased by ghosts. And all this peculiar behavior increased steadily as the weeks went by.

Toward the end of January a group of children went skating one afternoon on a local pond and were amazed to see frogs jumping through holes in the ice. Landing on the slippery surface, the frogs floundered helplessly until they were stilled by the freezing winds. The children cut short their skating expedition to hurry home and report these strange sights. On the way they found two snakes lying on top of the snow, their bodies frozen stiff. One of the boys picked up a frozen snake and carried it home to frighten his little sister.

About the same time, too, local zookeepers were observing unusual behavior in their animals. The tigers and lions paced restlessly in their cages. A young deer jumped too high and broke his leg. A panda held his head in his paws and moaned.

These strange occurrences were reported immediately to the commune committee that had been organized to observe and report any signs that might presage an impending earthquake. The committee passed the information along without delay to the Liaoning Province earthquake office, where data from the surrounding areas were analyzed. Many unusual phenomena had been observed in the past few weeks. The level of water in most wells had been rising dramatically. Some were delivering foaming, evil-smelling, foul-looking liquid. On one pond the ice had broken and warm, bubbling water had shot up like a fountain from an artesian well. The night sky over the Liaotung peninsula was occasionally illuminated with flickering orange lights, accompanied by booming sounds, as though a fireworks factory had exploded just beyond the horizon.

These mysterious portents were reported with greater frequency as the days passed. Like notes of music rising to a crescendo they seemed to presage an impending climax. Putting this information together with the data from many scientific measurements and the detection of a series of mild tremors, the local earthquake experts decided that a large-scale earthquake was imminent in the Haicheng area. This opinion was forwarded from the Liaoning Province earthquake office to the Provincial Committee just after midnight on February 4. The Provincial Committee warned all the provinces through emergency telephone calls and transmitted instructions to the personnel responsible for earthquake disaster prevention. The sick and elderly were moved to temporary shelters hastily erected outside the city. At 6:00 P.M. an announcement was made over the public-address system in the Kuan-t'un Commune: "According to a prediction by the superior command, a strong earthquake will probably occur tonight. We insist that all people leave their homes and all animals leave their stables." Movies were shown in the public square to encourage the people to stay outside, away from buildings. The first film was just finished at 7:36 P.M. when an earthquake registering 7.3 on the Richter scale struck Haicheng. Lightning flashed and thunder rolled and the earth rocked on waves of motion like a boat at sea. Ninety percent of the loosely constructed masonry buildings were demolished but only thirty people died and those were ones who had refused to heed the warnings. For the first time in the history of the world a large earthquake had been accurately predicted and the people led to safety.

~~

The expertise that made this remarkable accomplishment possible had been developed over the previous decade through the efforts of hundreds of thousands of people. In 1966 a strong earthquake had struck Hopeh Province, south of Haicheng, resulting in great loss of life and property. In the wake of this disaster the government of the People's Republic organized a massive national effort to understand and predict earthquakes.

Throughout their long recorded history the Chinese people have suffered terrible losses from earthquakes. Their country occupies a region of very high seismic activity and has a dense population living in brick, adobe, and tile-roofed buildings that are extremely susceptible to earthquake damage. So loss of life from this cause has been greater there than in all other countries combined. On the positive side China has several important advantages as a place to conduct seismic research. It has been the site of the longest consecutive civilization on earth; written records go back fourteen centuries before the Christian Era. The occurrence of an earthquake was documented as early as 1189 B.C. Under the People's Republic the population is closely organized and willing to undertake cooperative projects. Public participation is possible there to a degree that can hardly be matched anywhere else in the world.

For several years after the earthquake program was initiated teams of seismologists and scholars researched Chinese history for records of earthquakes and descriptions of the various phenomena that had been observed preceding these events. The vast research effort, conducted without the aid of computers, yielded significant evidence that unusual animal behavior, changes in water levels, illumination of the night sky had all been noted many times immediately preceding major shocks. There were even reports that well water had turned the rice red. A frequently repeated pattern was the occurrence of many medium-sized tremors followed by a quiescent period and then a series of small shocks coming in rapid succession just before the major quake itself. Changes in ground level had sometimes been noticed even as long as eight or ten years before the earthquake. Thus history reinforced folk wisdom. Certain unexplained phenomena seemed to foretell impending disaster.

Placing confidence in this conclusion, the Chinese government organized an enormous network of both amateurs and experts to collect and interpret any occurrences that might be construed as earthquake precursors. Citizens monitored water levels in the wells and watched for peculiar animal behavior. Trained seismologists took regular measurements of ground-level lines in areas known to be especially earthquake-prone.

At about this same time geologists in the Soviet Union were also making a concerted scientific effort to identify warning signals. In the late 1960s they announced several important discoveries. They had detected changes in the electrical resistivity of the earth preceding an earthquake. The amount of radon, a radioactive gas associated with groundwater, also increased. Furthermore, the speed of shock waves set up by earth tremors seemed to change in a manner that could be used to estimate both the timing and intensity of the quake.

When the crust of the earth heaves and cracks, vibrations are set up that travel in a complex pattern through the planet like the ringing of a great bell. Different kinds of waves travel at different speeds and are reflected or refracted in many ways as they enter and pass through the layers of the earth. By separating out the individual waves and studying how they have interacted, somewhat like analyzing the vibrations of music, seismologists have been able to deduce a great many things about the structure and composition of the planet.

The vibration that travels fastest away from the epicenter is called the pressure wave (the P wave). It causes alternating compression and dilation of the rocks in the plane in which the wave is traveling. The shear wave (the S wave) causes alternate compression and dilation in the plane perpendicular to the direction of its motion and it travels a little more slowly. P waves are like the disturbance that passes along a soft spring such as a Slinky toy. S waves are like the vibrations of a violin string when it is plucked. There are also several types of waves that are propagated only along the earth's surface. These have their own characteristic speeds, which are slower than the S wave. The difference in the arrival times of the P and S waves at any point on the earth's surface has long been used as a way of measuring the distance from the epicenter. This method depends upon the fact that the velocities of P and S waves are usually quite constant in the lithosphere.

The Russian scientists reported, however, that the P wave traveled more slowly than normal in the region where an earthquake was imminent. The speed of the S wave was not affected; so a decrease in the difference between P and S arrival times could be taken as a sign that an earthquake was likely to occur in the region. Furthermore, the duration of the deviation from normal was an indication of the intensity of the disturbance that could be expected.

All of the anomalies — increased electrical resistance, greater concentrations of radon, and decreased P wave velocity — suddenly returned to normal, the Russians said, just shortly before a quake occurred.

These discoveries fit quite neatly with information about the characteristics of rock subjected to great stress. When the applied pressure reaches approximately half the amount necessary to cause a major fracture, the rock begins to open up in many small cracks that actually cause it to expand in size. In this dilated condition it becomes more resistant to the passage of electricity, releases more of its content of gases, and transmits pressure waves less efficiently. As more stress is applied, water begins to flow into the cracks. Since water is a better conductor of electricity than rock, the electrical resistance drops as groundwater levels rise. These characteristics might explain why snakes comes out of their holes and fountains of bubbling water break the ice. As the rock is further weakened by saturation with water, the small earthquakes known as foreshocks occur and the main shock follows swiftly.

When these discoveries were first announced, they were greeted with great enthusiasm by seismologists. It looked as though earthquake prediction had at last been placed on a solid scientific footing. Within a few years however, conflicting evidence cast a shadow across this optimism. Measurements made in the United States failed to confirm consistently the Russian observations of changes in P wave velocity. The Chinese faith in abnormal animal behavior as a precursory sign was not shared by scientists in the West. Even if the Russian and Chinese methods were shown by subsequent testing to provide valid signals of an impending earthquake, there are problems to be overcome in gathering and interpreting the information.

Precise timing of the event is difficult or impossible in many cases. If precursory signs have been observed for, say, a dozen years, is the quake likely to occur next year, next month, or tomorrow? In some cases the return to normal of the rock anomalies may happen very suddenly and not give sufficient warning. Furthermore, a false alarm can have serious economic and political consequences.

The Chinese people have solved some of these problems by enlisting the cooperation of the whole population. Countless teams of "barefoot scientists" have been organized and 10,000 trained seismologists are working in 5,000 earthquake stations and 250 observatories around the country. This effective use of a large manpower base is one of the great strengths of the program. In their society there is no reason for concern about the economic or political consequences of a false alarm. The people feel that they are in partnership with the scientists, helping to mitigate one of the most terrible scourges of mankind. In spite of this vast cooperative effort, however, earthquakes can still take a Chinese city by surprise, as the one million inhabitants of the city of Tangshan discovered in the early morning hours of July 28, 1976.

Tangshan is an important manufacturing city in a very populous district of China about 300 miles from Haicheng as the crow flies. For several years before 1976 ominous changes in the magnetic field and other rock characteristics foretold that a quake might be expected in this area. However, none of the short-term precursors that usually announce a major quake had been observed. This time the earth gave no warning.

It was 3:40 A.M. in the darkest hours of a wild and windswept night when a devastating earthquake struck Tangshan like a thunderbolt hurled by Jove. Most of the people were in bed except for the men working the night shift in the coal mines directly below the city. Lightning flashed; a mighty rumbling noise came from the earth; buildings lunged into the air and tore apart. The town was almost 100 percent destroyed and the mines deep beneath the streets collapsed, entrapping the night shift. The first shock registered 8.2 on the Richter scale. It was followed sixteen hours later by another measuring 7.9. The Tangshan earthquake is believed to have been one of the world's worst earthquake disasters in recent centuries, but life has gone on. The city is now almost entirely rebuilt — of brick and adobe, tile-roofed buildings.

One of the most frightening effects associated with many earthquakes was not felt at Tangshan, which lies about thirty miles inland from the sea. When faulting occurs on the ocean floor or involves sudden movements of earth into the ocean, displacing substantial amounts of water, giant earthquake waves — tsunamis — are generated. These waves travel across the water at astonishing speeds (up to 500 miles an hour), but they are not threatening to ships in the open ocean. The waves are hardly noticeable there. The water level rises gradually a few feet in height and falls off again just as gently. However, as this extra volume of water approaches a shelving shore and in many cases is funneled by confining headlands flanking a bay or harbor, the waves pile up to enormous heights — ten, twenty, even a hundred feet above normal high-water mark. Tsunamis can strike with terrifying speed and force, descending on unsuspecting populations thousands of miles from the original quake.

Earthquakes have inspired fear and anger from the very earliest days of human history — fear because the cause was unknown, anger because the devastation was so inequitable. Some cities were destroyed over and over again while others were untouched. Why was this city singled out for destruction? Was it a punishment sent by the gods? Perhaps an omen of some greater disaster?

Plate tectonics can now provide some of the answers to these questions. We know that certain zones are particularly prone to earthquake activity and the majority of these zones lie on or near the plate boundaries. When the plates slide past or grind over each other, the frictional resistance to the motion can cause stress to build up until it reaches the point where the crust fractures and an earthquake occurs. This theory helps to identify places where earthquakes may be expected and where special vigilance should be maintained for the subtle signs of impending tremors (see Figures 2 and 5).

Mexico, for example, lies on the boundary between the Caribbean and the Cocos plates. The seismic belt stretches the length of the country and with limited resources available it is impractical to monitor the entire area; so ways must be found to pinpoint specific places. In 1975 geophysicists noted that for about two decades the region around Oaxaca had not experienced any of the moderate tremors (magnitude 4–6) that had been characteristic of this zone in the past. The period of quiescence might indicate that stress was building up — that it was not being relieved by many small earth movements. This possibility prompted the installation of a local seismic network, and fortunately it was in place early enough to record the flurry of small tremors that immediately preceded a major earthquake in 1978. This successful

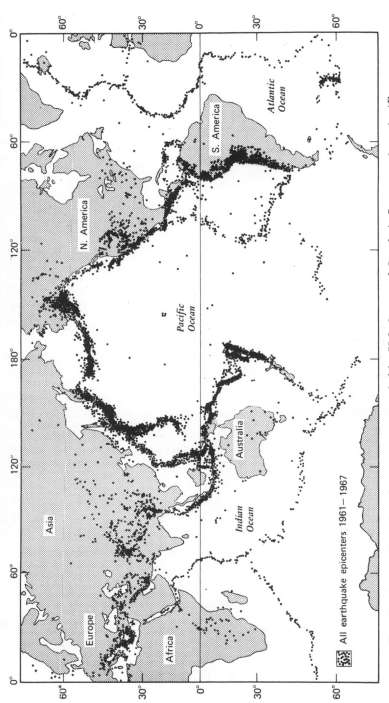

FIGURE 5 *Distribution of all earthquake epicenters recorded by U.S. Coast and Geodetic Survey, 1961–1967. (From M. Barazangi and J. Dorman, Bulletin of the Seismological Society of America 59, p. 369, 1969. Reprinted by permission)*

prediction and warning have inspired a search for other quiescent zones along the major fault lines of the earth.

But the earth has subtleties that science has not yet fathomed. Some of the most powerful earthquakes have occurred in places far removed from any plate boundaries. Tangshan and Haicheng, for example, lie well within the big Eurasian plate, nearly a thousand miles from its eastern edge and even farther from the place where the Indian plate is thought to be driving beneath Asia.

On the other side of the world three of the greatest earthquakes that have ever occurred within the United States in recorded history shook a region in the center of the North American plate. They struck in relatively rapid succession (within a two-month period in 1811) near New Madrid, Missouri, on the Mississippi River. Although the region was sparsely populated and few deaths resulted, the quakes had a profound effect on those who lived through the experience. Their reports speak of "sudden darkness caused by thick sulfurous vapor, the smell of brimstone, deafening explosions, and water, sand, and organic matter vomited to great heights." Many square miles of land were uplifted while others sank; the ground yawned in frightening fissures. Ponds, swamps, and lakes — one eighteen miles long — were suddenly created. Whole groves of young trees were snapped off like so many matchsticks. The Mississippi River bubbled and seethed. There were even reports of the current running backward. The ground rose and fell as waves passed across its surface. It continued to tremble for weeks at a time "like the flesh of a beef just killed." Many people suffered nausea, dizziness, and loss of balance. The shock waves of these earthquakes were felt from Canada to the Gulf Coast and east to the Atlantic seaboard.

Two hundred smaller quakes have been recorded at New Madrid during the past two centuries, but it was not until 1980 that geologists located the massive underground rock fault associated with this extraordinary seismic activity. They found shifts of 3,000 feet in the rock formations more than a mile below the surface "where the earth is actually tearing apart." But why has this great fault developed in the middle of the North American plate? Is it the remnant of an old plate boundary or a new one opening up? No one knows the answer.

Memphis is less than fifty miles south of this tremendous rift and, therefore, lies in a zone of high earthquake risk. Boston and Charleston,

South Carolina, have also suffered from major earthquakes in the past and, on that basis, are considered to be high-risk areas even though they also rest very comfortably within the North American plate, which extends from the Mid-Atlantic Ridge to the coast of California. It is a surprising fact that the peril may be as high in Memphis, Boston, and Charleston as it is in San Francisco near the famous San Andreas fault. In fact, one third of the population of the United States live in zones of high seismic risk.

The presence of the great San Andreas fault is much more easily understood in terms of plate tectonics than that of the fault at New Madrid. It lies right on the grinding edge between the North American and Pacific plates. Tremors are very frequent in the region and slippage along most of the fault can be measured year by year. A hundred-mile-long portion near the town of Palmdale, however, had not shown any slippage for approximately forty years while other portions had moved. About 1960 geologists detected what seemed to be a sudden uplift of approximately six inches extending over a 4,500-square-mile region. The "Palmdale pimple," as it has been called, has been watched apprehensively for the last two decades. During this time, however, scientific controversy has raged over the significance and accuracy of the changes in ground level. Some geologists maintain that the pimple does not exist at all, that the apparent elevation was the result of surveying errors. Others believe that the rise in ground level is real and should be considered an important warning sign that an earthquake could occur in the area.

Palmdale lies just forty miles northwest of Los Angeles with its seven million inhabitants. In case of a major quake the loss of life could be minimized if an accurate prediction were issued. But a false alarm, on the other hand, would cause social upheaval, economic loss, and probably a flurry of lawsuits from real estate speculators. Officials are not anxious to assume such a great responsibility unless they can make a prediction with certainty — an impossible requirement in the present state of the science.

In spite of these problems, considerable progress has been made in this century in dealing with so fearful a cataclysm of nature. We know in general where earthquakes are most likely to occur. Attention is being given to the art of designing buildings and cities that can withstand earth tremors. Perhaps the most important advance is understanding why the earth's crust trembles. Along with volcanoes and geysers and hot springs, earthquakes are part of the dynamic process that has formed our world and is even now continuing to create it. Spreading ocean floors and building mountain ranges, this activity produces the beautiful, varied landscapes of the earth and counterbalances the forces of denudation that would otherwise reduce this fascinating planet to a smooth and featureless globe. An understanding of the cause has helped to quiet the fear inspired by these events in ages past when an earthquake was believed to foretell an even greater catastrophe — perhaps the fall of an empire or the end of a dynasty. The physical damage of earthquakes was greatly enhanced by terror of the unknown and the predictions of doom often became self-fulfilling prophesies. There was at least one occasion when the occurrence of earthquakes and volcanoes preceded the fall of a great civilization. Geologic evidence supports the conclusion that these were not just harbingers of disaster but actually the cause, altering the course of human history.

Between the twentieth and the fifteenth centuries B.C. a highly refined and artistic civilization evolved on the islands of Crete and Santorini in the Aegean Sea. It was a seagoing culture depending on a large fleet of merchant vessels to carry on a lucrative trade throughout the Mediterranean and on the supremacy of this fleet to protect the islands from invasion. There were luxurious cities and splendid palaces — notably Knossos and Phaistos and Zakros on Crete and Akrotiri on Santorini. This latter island, sometimes called Thera, now lies in a group of five small islands, but in the early fifteenth century there was a single island here named Stronghyli (meaning round).

In this lighthearted and relatively peace-loving civilization the cities and palaces were not fortified. They contained many comforts and conveniences, such as toilets with flushing arrangements, drainage and sewage systems. Even the smaller dwellings were spacious, two or three stories high with courtyards and large windows. Writing (in a script that has not yet been satisfactorily deciphered), sculpture, painting, and exquisite pottery work flourished in the Minoan civilization, which preceded the heyday of Greece and was contemporaneous with the time when the Egyptian civilization was in full flower. But about the middle of the fifteenth century the Minoan culture went into a precipitous decline and within the space of one generation had passed from the front pages of history; it dropped from a position of great power and prominence to the obscurity of the back pages.

The reasons for this fall from glory have long been a subject of speculation. Could it have been caused by invasion, disease, rebellion, drought, decadence? All these possibilities have been considered and rejected. There are many signs of earthquake damage, but this has been common throughout the Mediterranean, which lies on the boundary between the Eurasian and African plates. The cities were usually rebuilt within a few years.

Recent information from deep-sea cores, microscopic analysis of soil, and studies of the geology of Santorini, combined with detailed archaeological evidence, now strongly supports the theory that the fall of the Minoan civilization was the result of a great volcanic eruption accompanied by tsunamis and earthquakes. The imaginative reconstruction that follows (proposed by geologist Dorothy Vitaliano) fits the facts as they are known today.

~⌒~

Sometime early in the fifteenth century B.C. the inhabitants of the Minoan city of Akrotiri on Stronghyli noticed a series of mild earth tremors. Since earthquakes were a common experience, no one was very much concerned at first. But the strength and frequency of the quakes increased and soon it became apparent that something unusual and alarming was happening. Shepherd boys tending flocks of goats high up on the mountain reported that the ground had become hot in several places. Steam and sulfurous gases were trailing in little plumes from cracks in the earth. Perhaps, some people suggested, Stronghyli was about to become a mountain of fire like those they had seen in Sicily and off the coast of Italy. True, their mountain had never erupted within human memory, but it did have a shape like a volcanic cone. As the frequency and strength of the tremors increased, the populace became more apprehensive. A few of them packed up and left. (The inhabitants of Stronghyli were more fortunate than those of St. Pierre, who were dissuaded from leaving their city under similar circumstances.)

One night flames erupted high on the flank where hot gases had been issuing from fissures in the ground. Panic broke out in the city and the remaining inhabitants hastily gathered their most precious objects and crowded onto the boats in the harbor. The city was entirely evacuated except for several old people who refused to leave. The last boats had cleared the harbor when the mountaintop burst open with a tremendous roar. Steam and ash rose in a great plume and spread out, making a gigantic mushroom cloud that cast a dark shadow over the surrounding sea. Violent explosions followed each other every few minutes, growing in intensity. As night fell, flaming rocks and pieces of tephra traced fiery trajectories high in the air and came back to rest on the lower slopes of the mountain. Pieces of pumice fell and shattered in the empty city streets.

Over a period of several days or weeks the eruptions continued intermittently, alternating with short quiescent periods. Then suddenly the tempo began to accelerate. When the climax came, it was more violent than any witnessed by man during recorded history. Explosions were heard all the way to Scandinavia and far into Asia and Africa. Shock waves damaged buildings throughout the Aegean Islands. Pumice covered the ocean surface for miles around, so thick that a man might walk on it and no ship could pass through it. But the most awesome effect was the total darkness that came as the ash cloud descended on the neighboring lands. It was a darkness like sudden blindness. No ray of light crept through. No candleflame could relieve the absolute blackness.

When the darkness finally lifted on Crete, just sixty miles south of Stronghyli, a grayish white layer of ash could be seen lying like a shroud over the olive orchards and fields and vineyards throughout most of the eastern half of the island. The crops were smothered under the volcanic dust and although heavy out-of-season rains, precipitated by the extra particulates in the air, drenched the area and rinsed ash off the steep hillsides, the lower land where most of the crops were raised was buried even deeper as the ash was washed down from higher ground. Oozing mudflows filled the valleys. With widespread crop failure the Cretans suffered great privation that winter.

The eruptions subsided and time passed. Then suddenly one day the sea withdrew from the little harbors and beaches along the northern shores of Crete, retreating to levels far below that of the lowest tides. Boats that had been anchored in the harbors were stranded on mud flats. Rocks and crustaceans that had never felt the sea breeze before were laid bare. As though the earth had drawn a deep breath and then abruptly exhaled, the water returned in a solid wall. It swept landward and smashed into the shore with tremendous force, flooding the little villages in its path. Then, its fury spent, the water slowly drained back to the ocean, carrying with it everything that had been knocked loose — uprooted trees, houses lifted from their foundations, and people — out to sea.

The older and more experienced inhabitants remarked that the terrifying wave was just like waves that sometimes follow earthquakes. But no violent tremors had warned of the imminent danger. Similar visitations of death from the sea occurred frequently in the next few years. The destruction of ships and harbor facilities was a severe blow to this maritime nation.

Sailors who passed by Stronghyli noticed that the old round island had now become several smaller islands and the shape of these was changing all the time. The center of the volcano had sunk below sea level. The caldera rim had been breached and the sea had rushed in, making a large bay where the top of the mountain had once stood. (This sudden change in sea level had precipitated the first tsunami wave, but the Minoans did not understand this relationship.) Every little while more pieces of the caldera collapsed and fell into the bay, causing the waves that continued to harass the coast of Crete. The Cretans who lived on higher ground back from the seashore were not directly affected except that these unexplained natural disasters increased their anxiety. No one understood the cause of the great waves that struck without warning, and fear grew in the hearts of the people. It was apparent that the gods had singled out the Minoans on Crete and Stronghyli for some special punishment.

Then one night the gods lashed out again with decisive fury. A major earthquake centered deep under the city of Zakros in eastern Crete shook the entire island. The mud-brick multistory buildings rocked and toppled like houses built of toy blocks. Oil lamps that just a few minutes earlier had cast their mellow glow on rich banquet scenes and luxurious interiors were overturned. The holocaust of fire was added to the din of collapsing masonry and the screams of victims caught in its path. In Zakros the fire was so intense that large pieces of

adobe brick were partially melted. Many of the other major towns in
Crete were severely damaged.

Survivors of this disaster did not have the heart to rebuild as their
fathers and grandfathers had done so many times before them. It was
apparent to these well-traveled people that Crete and Stronghyli had
been afflicted as no other places on earth had been within living
memory. Better to leave this cursed land and go on to other places
that were in higher favor with the gods. Those who had the means
picked up what was left of their belongings and moved away. The
Minoan nation, depleted of its most able citizens, never recovered its
strength and power. Many of the refugees settled in Mycenae on the
mainland of Greece, bringing with them a high cultural tradition that
contributed to the emergence of the Greek civilization and through it
to the whole Western world.

Another exodus that had profound implications for human history
may have been precipitated by this same awesome display of natural
forces in the middle of the fifteenth century B.C. The powerful and
important land of Egypt lay roughly five hundred miles south of
Santorini, directly in the path of the prevailing northwesterly winds.
Many of the alarming manifestations suffered by the Minoans must also
have been experienced on a somewhat lesser scale in Egypt.

Very few contemporary records have survived from this period of
Egyptian history, which fell in the Eighteenth Dynasty. However, later
writings refer to historical events believed to have taken place at this
time. From the Hermitage papyrus in Leningrad comes this description:

> The sun is veiled and shines not in the sight of man.
> None can live when the sun is veiled by clouds.
> None knoweth that midday is here . . . his shadow is not
> discerned. Not dazzled is the sight when he [the sun] is
> beheld . . . he is in the sky like the moon.
> The river is dry, even the river of Egypt.
> The south wind shall blow against the north wind.
> The Earth is fallen into misery.
> This land shall be in perturbation.
> I show thee a land upside down: that occurred which never
> yet had occurred before.

And from the Ipuwer papyrus in the Leiden museum:

> Plague is throughout the land.
> Blood is everywhere. The river is blood. Men shrink from
> tasting . . . and thirst after water.
> All is ruin! Forsooth gates, columns and walls are
> consumed by fire. The towns are destroyed. Oh, that the Earth
> would cease from noise, and tumult be no more.
> Trees are destroyed. No fruit or herbs are found . . . grain
> has perished on every side.
> All animals, their hearts weep . . . cattle moan. Behold,
> cattle are left to stray, and there is none to gather them
> together.
> The land is without light.

It has been suggested that these accounts, strongly reminiscent of earthquake effects, are also similar to the biblical story of the plagues that descended upon Egypt in the time of Moses and persuaded the Pharaoh to allow the enslaved tribes of Israel to go into the wilderness and worship their God according to their own customs. Ten punishments were inflicted on Egypt, as described in Exodus:

1. ". . . all the waters that were in the river were turned to blood. And the fish that was in the river died; and the river stank, and the Egyptians could not drink of the water of the river; and there was blood throughout all the land of Egypt" (7:20–21).
2. ". . . and the frogs came up, and covered the land of Egypt" (8:6).
3. ". . . all the dust of land became lice throughout all the land of Egypt" (8:17).
4. ". . . the land was corrupted by reason of the swarm of flies" (8:24).
5. ". . . and all the cattle of Egypt died" (9:6).
6. ". . . they took ashes of the furnace, and stood before Pharaoh; and Moses sprinkled it up toward heaven; and it became a boil breaking forth with blains upon man, and upon beast" (9:10).
7. ". . . and the Lord sent thunder and hail, and the fire ran along upon the ground; and the Lord rained hail upon the land of Egypt . . . the hail smote every herb of the field, and brake every tree of the field" (9:23–25).
8. ". . . and the Lord brought an east wind upon the land all that day,

and all that night; and when it was morning, the east wind brought the locusts. . . . They covered the face of the whole earth, so that the land was darkened; and they did eat every herb of the land, and all the fruit of the trees which the hail had left: and there remained not any green thing in the trees, or in the herbs of the field, through all the land of Egypt" (10:13–15).

9. ". . . and there was a thick darkness in all the land of Egypt three days" (10:22).

10. "And it came to pass, that at midnight the Lord smote all the firstborn in the land of Egypt, from the firstborn of Pharaoh that sat on his throne unto the firstborn of the captive that was in the dungeon; and all the firstborn of cattle . . . and there was a great cry in Egypt; for there was not a house where there was not one dead" (12:29–30).

Many of these plagues can be interpreted as major volcanic and earthquake phenomena: the sudden darkness, the death of fish and livestock, the thunder, fire, and hail. Plagues of desert locusts have been observed to occur in the wake of heavy rains, which cause a prolific growth of desert vegetation. Several of the phenomena are more difficult to explain in this context but they carry haunting reverberations of reports we have heard from other times and other parts of the world. (In China the frogs jumped out of the ponds and rivers; the water of the wells turned the rice red, the waters frothed and were evil-smelling, the panda moaned and held his head in his paws.)

Geologists have suggested that the dust which caused sores on the skin of men and beasts might have been volcanic ash contaminated by traces of fluorine. This is a phenomenon that has been noted in several volcanic eruptions. Fluorine is poisonous and corrosive, resulting in the death of livestock that graze on grass dusted with the ash.

Widespread sickness and death have frequently occurred following major earthquakes. Hygiene breaks down. In the abnormally damp weather insects breed more rapidly, helping to spread disease and destroy crops.

Several of the plagues, such as the smiting of the firstborn, defy explanation on this basis, but we should remember that the biblical story was written down several centuries after the event. A dramatic saga told and retold over many generations, its details may have been altered in the telling.

The most striking correlation with earthquake phenomena is found in the miracle of the parting of the "Red" Sea. Scholars have long held that the body of water described in the Bible was not the Red Sea but the Reed Sea — a literal translation of *Jam Suf* in the Hebrew original. Since reeds do not grow in salt water, the sea traversed by the fleeing tribes must have been a shallow lake or lagoon similar to the ones that lie on the border between the Nile delta and the sea. These lagoons would have been dramatically affected by a tsunami descending across the Mediterranean from the north. As the wave approached, the water would be drawn back into the sea, baring the bottom of the lake and allowing a passage across it — "into the midst of the sea upon the dry ground," as the Bible says (Exodus 14:22). Then, possibly as long as half an hour later, the waters would return in a great wall, drowning the forces that the Pharaoh had sent to overtake them. "And the waters returned, and covered the chariots, and the horsemen, and all the host of Pharaoh that came into the sea after them; there remained not so much as one of them" (Exodus 14:28–29).

The timing of the Exodus has never been firmly established by biblical scholars. Some believe it occurred about 1200 B.C., but others think that it was earlier, perhaps in the middle of the fifteenth century. The date 1446 has been proposed, based on a statement in Kings that Solomon started to erect the temple of Jerusalem in the fourth year of his reign, 480 years after Exodus. Solomon's reign has been dated with a reasonable degree of accuracy as beginning in 970 B.C. On the other hand, evidence cited by Egyptologist Hans Goedicke suggests a slightly earlier date. Inscribed on a rock above the entrance to a shrine of the goddess Pakht at Speos Artemidos is a royal decree of Hatshepsut, who reigned between 1490 and 1468 B.C.:

> While I restored what had decayed, I annulled the former privileges [that existed] since [the time] the Asiatics were in the region of Avaris of Lower Egypt! The immigrants among them disregarded the tasks which were assigned to them, thinking Ré would not consent when the deified [i.e., her father, Thutmosis I] assigned the rulership to my majesty! When I was established over on the thrones of Ré, I became known through a period of three years as a born conqueror. And when I came as king, my uraeus [symbol of royal power] threw fire against my enemies!

And when I allowed the abominations of the gods to depart, the earth swallowed their footsteps! This was the directive of the Primeval Father [i.e., Nun, the primeval water], who came one day unexpectedly.

If we accept the conclusion that the inscription refers to the Exodus of the Israelites from Egypt, then we should ascribe to it a date of about 1477 B.C. This date is in close agreement with the most recent estimates of the eruption of Santorini and the fall of the Minoan culture.

❧

Today the little island of Santorini dreams in the sunshine. The whitewashed houses of Phira are set like a crown of pearls high above the azure blue of the Aegean Sea. Below the town precipitous cliffs descend to the bay where once the summit of Stronghyli rose. Motorized vehicles have difficulty negotiating the switchbacks on the steep trail that zigzags up the mountainside so the ascent to the town can best be made on donkeyback. The slow pace and hypnotic rhythm of the donkey's footsteps impart a timeless quality to this trip across the great caldera rim with ancient history written on its fractured face. Layer upon layer of ash and lava and pumice lie quietly folded now, pressed together like flowers in the pages of a book. Only the striking spectrum of colors — deep red, rose, pale pink, chocolate, light buff — suggests the drama recorded here. In this bright dust is the story of days and nights of terror: cities deserted, armies drowned, navies crushed, and the whole course of human culture redirected down new paths.

The earth here is at rest, but it will move again. It will tremble beneath the blue Aegean, beneath the sand dunes of Death Valley, and the cherry orchards of Japan. These events are inevitable considering the vitality, the strength, and the power embodied in the earth — this massive ball of matter swinging in its orbit around the sun a hundred times faster than the speed of sound and spinning through space ten times more rapidly still on the vast pinwheel of the Milky Way. The energy involved is almost past imagining. Seen in the perspective of the whole life and development of the planet, a major earthquake is a little thing, a mere twitch of the skin to relieve a small discomfort just below the surface.

And man — man in his tiny adobe dwelling perched on the edge of the precipice — will be shattered again and again by these tremors until he learns to understand the earth, to sense its stirrings, to interpret the past and foretell the future of this deceptively "quiet dust" that he can hold cupped in the hollow of his hand. " 'Tis natural," the poet John Wheelock tells us,

> *Yet hardly do I understand —*
> *Here in the hollow of my hand*
> *A bit of God Himself I keep,*
> *Between two vigils fallen asleep.*

7

How Ice Changed the World

> The afternoon may be so clear that you dare not make a
> sound, lest it fall in pieces. And on such a day I have
> seen the sky shatter like a broken goblet, and dissolve
> into iridescent tipsy fragments — ice crystals falling
> across the face of the sun.
>
> — RICHARD E. BYRD, *Alone*

The long sensuous days of summer in the temperate zone of the Northern Hemisphere are past as the planet Earth swings relentlessly in its orbit. The faces of the lands we know turn away from the sun a little farther each day. The shadows of the night lengthen week by week and the winds spin out of the north. One day the first snowflakes fall softly from the darkening sky.

On the lakes and ponds a thin film of ice crystals makes a translucent skin, hardly visible to the naked eye but changing the way the surface turns back the light. The wake of the moon is diffused and softened. The water's restless movement is quieted; its body rises and falls almost imperceptibly as though it had fallen asleep.

The days pass and the ice thickens, clearly visible now in thin sheets and disks that nudge against each other. As they move and grow, they acquire rounded shapes with little upturned edges — "pancakes," these ice platelets are called, but they are more like shallow saucers. Along the shore they pile up and slowly hour by hour build fantastic castles with high walls and moats and turrets and battlements and deep wells down through the ice. Into these holes waves send sudden jets of water that spout fountains high into the air. The drops freeze as they fall in a sparkling shower of crystals, diamond-bright.

Along the drifted country roads, beside the shining seashores, through the softened spaces of forests and cities dressed in white, for a brief period of time frozen water has redesigned our world. But in much more important and fundamental ways ice has been creatively at work since the crust of the earth first began to form four billion years ago.

❧

The contours, the color, and texture of all the land surfaces have been affected by ice, because without ice we would not have rain, which distributes the moisture across the continents, erodes and transforms the hard rocks of the continents into friable soil, provides a universal solvent, and distributes the most essential ingredients for living things. Raindrops do not usually form directly from clouds. Water passes first through the frozen state; raindrops are melted crystals of ice.

When water evaporates from the surface of the seas or lakes, it becomes a gas, composed of single molecules free to move independently among the other molecules of the atmosphere. Warm air can hold more water in the vapor form than can air at lower temperatures. When the atmosphere cools, the water molecules tend to condense back again into tiny cloud droplets many times smaller than raindrops — a billion cloud droplets would not fill the hollow of your hand. These tiny drops of water are the substance of all the clouds that veil the sky, the mists that rise from marshlands on a summer night, the dew that collects on meadow grass at dawn. But they are at least a million times smaller than raindrops. They are so light that they remain airborne, suspended by the rapidly dancing molecules of the gases in the atmosphere, and they do not tend to come together to form larger drops. So for many decades meteorologists were baffled in their attempt to explain how raindrops form. Then it was discovered that a frozen cloud droplet is the seed that makes raindrops. A growing ice crystal attracts water-vapor molecules, persuading them to give up the individuality they had preserved so carefully before the crystal appeared. As each molecule is added to the crystal it draws in others. Like an elaborate circle dance, an intricate, symmetrical pattern is built up in ever-widening spheres. A unique, original masterpiece of form suddenly materializes from what appeared to be empty space.

Soon the ice crystal becomes heavy enough to fall through the cloud and as it collides with cloud droplets, little splinters of ice break off, leaving a trail of tiny fragments that act as nuclei for new crystals. These also grow and fall and splinter. In the space of just a minute or two the cloud becomes a flurry of snowflakes. As they descend into warmer air they melt and descend as rain upon the land — rain to fill the lake basins and swell the rushing rivers, to carve out canyons, and soften the harsh rock surfaces, to tear down the mountain ranges stone by stone and return their substance to the sea.

In wintertime, of course, the snow crystals do not melt. They swirl gently downward, collecting in a downy blanket on the ground, covering every twig and rooftop. Freshly fallen snow is incredibly airy and light. A typical powder snow has twenty parts of air to just one of water. But as it remains on the ground, it gradually becomes compacted into a solid mass. Packed snow melts very slowly even when the summer sun shines on it. If the season is short and cold, the snow does not all melt away. It accumulates year after year. Thus a glacier is formed.

Many glaciers exist today at high altitudes, creating breathtaking mountain scenery. The ice collects in bowl-shaped depressions near the mountain peaks where the weight and movement of the growing ice mass carve steep-walled basins called cirques. Eventually the ice overflows the basin and moves downhill, becoming a mighty river of ice, and great ice falls are formed where it cascades over precipitous cliffs.

As the glacier moves, it scoops out depressions that later, after the ice has melted, may become filled with water, making glacial lakes, or "tarns." Along mountainous seacoasts in the high latitudes glaciers have cut troughs deep down into the continental crust below present sea level, creating the beautiful fjord landscapes where the sea runs in shining threads far inland between towering rock walls.

In the polar regions extensive ice sheets cover the land and much of the water surface. Greenland and the Antarctic continent are almost entirely buried beneath glaciers measuring as much as two miles in thickness. The Antarctic ice sheet alone covers one and a half times the area of the forty-eight contiguous United States. In some places it extends beyond the borders of the continent, attached to the land but riding out over the sea.

Travelers to these frozen parts of the world have been surprised to find that the great ice masses are not all white or shades of gray. Even in the frozen state water retains its characteristic color, diluted to a faint gleaming aqua in the vertical walls of the great glaciers but intensified in the shadowy crevasses to a midnight blue. When seen from above, the icebergs that float free are a deep opaque tone like lapis lazuli. Admiral Byrd described his first impression of the remarkable ice sheet that he called "the Barrier" on his pioneering trip to Antarctica in 1928: "Near the water's edge, the Barrier in places was honeycombed with caves, of bewildering shapes and sizes which, when the sun struck them at just the right angle, blazed with a rich blue coloration."

Several months later Byrd flew from Little America over Marie Byrd Land, enjoying a really sweeping view of this tremendous continental glacier:

> There was great beauty here, in a way that things which are also terrible can be beautiful. Glancing to the right, one had the feeling of observing the twilight of an eternity. Over the water and submerged land crept huge tongues of solid ice and snow, ploughing into the outer fringe of shelf ice and accomplishing wide destruction. There were cliffs that must rise hundreds of feet. Once I caught sight of a cliff as it fell into the sea. From the great height of the plane it was just a small pellet falling from a toy wall. Not a sound penetrated through the noise of the engines. Yet thousands of tons must have collapsed in one frightful convulsion. To the right were the mountains, cold and

gray, and from them fell, in places, ice falls which were perhaps 500 feet in height. . . .

Here was the ice age in its chill flood tide. Here was a continent throttled and overwhelmed. Here was the lifeless waste born of one of the greatest periods of refrigeration that the earth has ever known. Seeing it, one could scarcely believe that the Antarctic was once a warm and fertile climate, with its own plants and trees of respectable size.

This mighty field of ice is only a small remnant of the last glaciation. Ice sheets have covered much larger portions of the planet at many times in the past. Just 18,000 years ago the lands that now support the streets and tall buildings of New York, Berlin, Stockholm, and Warsaw were bent under a heavy burden of ice. Frequent storms raged as they do now in Antarctica and winds lashed with a force and fury unknown on earth today. The story of these great ice ages can be reconstructed from the marks they left on the land held so long in their frozen grip.

As ice sheets recede they reveal characteristic impressions. There are grooves and striations on bedrock, making roughly parallel scars, like clawmarks on a treetrunk. These are sure signs that a glacier has pulled sharp rocks across this surface as it moved slowly forward. Low, rounded hillocks of assorted rock and stone occurring in gently curving arcs mark the places where the glacier's advance was halted and some of the rock debris which it carried with it from faraway places was deposited. These glacial moraines border outwash plains where finer material, like sand and gravel, also foreign in origin, was spread by meltwater streams in sheets before the leading edge of the ice. The plains are often pockmarked with small depressions known as kettles where buried ice blocks melted. Now these holes form little lakes and ponds. Here and there long, narrow ridges of coarse sands and gravel can be seen extending in sinuous courses for miles. This material was carried by streams that emerged from tunnels at the ice margins.

The amount of clay, soil, sand, and rock moved by continental glaciers staggers the imagination. Across hundreds of thousands of square miles the glacial drift was deposited and accumulated to great thicknesses as the ice melted. It covered the valleys, the canyons, and lowlands more deeply than the hilltops and plateaus, so the broadscale effect of retreating ice sheets was to smooth out the landforms and create great plains. In the United States land extending from eastern

Ohio through South Dakota was already relatively flat because it had been laid down in shallow seas, but it was given an additional smoothing by ice sheets several times during the last two million years. Glacial drift lies 50 to 200 feet thick over Iowa and Illinois. On top of the glacial drift is a layer of wonderfully fertile soil, which was also a gift of the ice age.

During the major glaciations winds were accelerated by extreme temperature differences on the planet. They blew across the ice sheets, picking up sand and dust from the alluvial valleys and outwash plains on the melting edge of the ice and distributing it across adjacent lands. Gradually this loess was built up into thick blankets extending over large portions of the continents. Loess is a fine-grained sediment, easily cultivated, and has a varied mineral composition, providing the important nutrients for plant growth. It is especially favorable for grain crops like corn and wheat, and it forms the parent soil for some of the richest agricultural lands on earth. Soft brown soil evolved from loess lies in a deep layer across the plain states of America and European Russia. In northern China loess accumulations sometimes reach thicknesses as great as 300 feet.

While the ice sheets were putting the finishing touches on the most fertile breadbaskets of the world, they were also busy carving out another important geographical feature — the Great Lakes. Before the last ice epoch this part of North America was a wide lowland area that had been severely eroded by streams. The heavy mass of the great advancing ice sheets scoured this lowland, deepened the basins, and altered the drainage patterns. As successive glaciations advanced and retreated, the places underlaid by the weakest strata subsided, creating enormous shallow bowls that soon became filled with water. The courses of the rivers and the shapes of the lakes changed many times during this period, finally culminating in the five Great Lakes that we know today.

❧

Ice has left its mark almost everywhere on the land surfaces, even in places that now lie on the equator and in the scorching sands of the deserts. In the early 1960s French geologists exploring for oil in the central Sahara came upon rocks scarred with glacial striations and grooves. They were working in the Hoggar region of southern Algeria, one of the most forbidding places on earth. A wilderness of jagged rocks and sands, it is blistering hot on summer days: temperatures of 137° F in the shade have been recorded. At these times no ant or snake or even lizard stirs until dusk. And the Arabs travel across the desert only after the burning sun has left the sky.

The age of the grooved rocks discovered by the French geologists could be estimated by distinctive fossils found embedded in the layers of rock directly above and below the strata bearing the glacial marks. They proved to be latest Ordovician, approximately 440 million years old.

Throughout the same northwestern region of Africa geophysicists had recently been taking measurements of magnetism in ancient rocks, and the results were quite surprising. Rocks dated between 400 and 500 million years old bore evidence that the South Pole had been located between the Cape Verde Islands and Morocco — a region that now lies on the Tropic of Cancer 23 degrees north of the equator. The equator at that time seems to have run diagonally across northern Europe, eastern North America, and Mexico.

This information suggested that a polar ice sheet had covered what is now the Sahara Desert 440 million years ago, but the French geologists were cautious about drawing this conclusion. Striations can be made by mountain glaciers as well as by polar sheets. Perhaps a great mountain range had occupied this part of North Africa. It was necessary to obtain more information before the question could be settled.

In 1970 a group of specialists undertook an expedition to study the evidence. In the Hoggar region and the Tassili Plateau a little farther east they found very ancient rocks, 600 million to two billion years old, bearing unmistakable evidence that warm shallow seas had covered this land and a tropical climate had prevailed. Ripple marks were found and the tracks of trilobites, a form of crablike creature that was ubiquitous throughout the shallow seas during this very early period of earth history. Ordovician sandstones overlay these most ancient rocks and in this stratum changes in environmental conditions were clearly recorded. The rocks at the base of the layer were typical of desert conditions, showing abrasion by blowing sand. At this time, then, the ancient seas must have retreated from the land. But higher up in the sandstone layer evidence of shallow seas and the tracks of trilobites were present again, as well as impressions of seaweeds and sand waves characteristic of low tide along coastal waters.

In some areas very deep channels had been cut by tremendous currents sweeping over the seafloor. Giant waves, ten feet from crest to crest, had been impressed on the rocks. Many of these great channels were overlaid with deposits of ill-sorted sands, pebbles, and boulders obviously of foreign origin. They must have been transported from considerable distances and were too heavy to have been carried by ocean currents or streams, but they could have been moved by ice. The deep channels might have been cut in the rocks, the geologists theorized, when tremendous volumes of water, released by melting glaciers, rushed back to the sea. Long sinuous ridges like the rock deposits of meltwater streams wound for miles across the landscape and were especially spectacular from the air. The rocks formed from these sediments had weathered out, cracked, and toppled into jumbled piles of bizarre shapes — forests of giant mushrooms, dark angular needles and spires.

These findings all bore testimony to glacial action on a tremendous scale, but the most striking evidence was found in the eastern part of the Hoggar in the valley of Wadi Tafassasset. Here mile after mile of parallel striations had scarred the rocks as though a great comb had been dragged across the land from southeast to northwest. From the air these scars can be followed for hundreds of miles. The conclusion was inescapable, even for the most cautious: a polar ice sheet had lain across the land that is now one of the hottest and driest places on earth 440 million years ago.

❧

Five glacial epochs (major ice intervals during which briefer ice ages may wax and wane) have been identified in the geologic record. Evidence of the earliest one is widespread in eastern Canada, where rocks 2.2 billion years old are marked with glacial striations and contain an unsorted jumble of large boulders, pebbles, and sandlike till dropped from melting ice. This period of extreme cold occurred long before life began to burgeon on earth. The next epoch of which we have any knowledge took place between 700 and 600 million years ago, when only soft-bodied organisms were present in the sea. The history of this Precambrian glaciation is fragmentary, but it seems to have involved Australia, South Africa, China, Europe, and North America. In fact, the entire world may have been covered with a seal of ice. The glaciers were not just polar phenomena; they even affected areas that lay on the equator at that time.

After this ice epoch the world passed through a long period when there were no glaciers or polar caps. About 200 million years after the Precambrian glaciation the Ordovician ice sheet formed in Africa near the South Pole of that period. By Silurian times, a few million years later, the ice had disappeared and tropical conditions had returned to the earth. Again there was a long warm interval — perhaps 150 million years — before the next ice epoch descended. (There is some evidence of glaciers in South America about 350 million B.P., but it has not been proved that a continental ice mass was responsible. They may have been just mountain glaciers.) In late Carboniferous time — about 290 million B.P. — ice sheets formed in Antarctica, which was near the South Pole at that time also. As the plates drifted around the world, other continents moved into the polar region — South America, Africa, India, and finally Australia. Glaciations appeared in each of these areas successively and did not entirely disappear from the planet until about 270 million years ago. Once more the earth was without large ice masses for a very long time until 2 to 3 million B.P., when the last ice epoch began, the glaciation that has had a profound effect on man and is still a dominant feature of the earth's climate.

Evidence of the most recent ice epoch is relatively fresh, so it is not surprising that these glacial remains were the first to be identified. They told a dramatic story of successive advances and retreats across the continents. For many years, however, the number and complexity of these Pleistocene glaciations were grossly underestimated. Until the late 1950s it was universally assumed that four advances to moderate latitudes had occurred in the last two million years. Only after the development of several sophisticated techniques for studying ancient climate and temperature conditions was the more complex history revealed.

Cores obtained by deep-sea drilling contain shells of small marine animals. Under a microscope paleontologists can count the numbers of fossil species whose range is markedly affected by climate. Using these counts, they can estimate to within two or three degrees the temperature of the water in which these organisms lived. Another measure utilizes the relative abundances of two isotopes of oxygen in the fossil shells: ordinary oxygen 16 and the heavier isotope, oxygen 18. When water evaporates from the ocean, water molecules containing the lighter isotope leave the surface more easily. So water with a higher proportion of oxygen 16 to oxygen 18 becomes cloud droplets and then the snow that creates the great glaciers. A larger proportion of oxygen 18 is left behind in the ocean water. This changing ratio is recorded in the shells of marine organisms.

Using these techniques for analyzing seafloor cores, geologists found a much more rapid succession of recent ice ages than had been identified in the glacial markings on terrestrial rocks. It appears that there may have been as many as twenty major glaciations in the last two million years. The cold periods have lasted much longer than the warm ones. Glacial periods averaging about 100,000 years in length have actually dominated the earth's climate for the whole period of time that human beings have occupied the planet. The warm interglacial periods have been only brief respites from the cold — typically 10,000 years in duration — and the one we have been enjoying now for about 9,000 years may have nearly run its course. However, in order to make any sensible prediction about the return of an ice age we must understand their cause. This is a subject on which there are many theories and no clear consensus today.

❧

Solar energy is the main driving force of the earth's climate, and therefore any changes in the luminosity of the sun could have an effect on the earth's climate. The amount of solar energy falling on one square centimeter at normal incidence measured outside the earth's atmosphere is known as the solar constant — a term implying an immutable property of nature. But actually there is no logical reason to assume that the sun must continue to produce exactly the same amount of energy year after year and eon after eon. A close examination of the solar surface has shown vast disturbances, like tremendous storms, that wax and wane, appear and disappear in a cyclical manner throughout a period that averages approximately eleven years. Although the hypothesis has not been proved, there is some evidence suggesting a causal relationship between the sunspot cycle and certain weather phenomena on earth: the occurrence of droughts on the Great Plains every other solar minimum and a cyclical variation in growing conditions revealed by tree-ring studies. Scientists at the National Center for Atmospheric Research have demonstrated a striking correlation between longer-term anomalies in solar activity and prolonged periods of abnormal climate. The seventy years between 1645 and 1715 were a time when the usual manifestations of solar activity such as sunspots and aurora borealis all but stopped. This period of reduced solar activity corresponded with the coldest years of the period known as the Little Ice Age, a time of unusually severe weather in Europe from 1600 to 1850. Modern climate models show that a reduction in global temperature sufficient to produce the Little Ice Age could be brought about by a decrease of one or two percent in total solar radiation. These facts suggest that solar radiation has not been constant throughout earth history; on the other hand, progressive changes as large as one or two percent in several centuries could not have been sustained for any considerable portion of geologic time without producing dramatic changes that would be obvious in the rock record. Variations of this magnitude, if they do occur, must be periodic or fluctuating.

However, it has recently been pointed out by astronomers that a continuous change of a much lower order of magnitude probably did occur. Stars like our sun pass through an evolutionary development, increasing gradually in luminosity as they grow older. This is one of the most widely accepted conclusions drawn from modern astronomical theory. The estimates for the sun vary between 30 and 60 percent increase in luminosity over the last five billion years.

Taking the most conservative figure of 30 percent and running the evolution of the sun backward through time, we find the surface temperature of the earth would have been below the freezing point of seawater about 2.3 billion years ago. Liquid water could not have existed anywhere on the planet. But this result is in strong contradiction to the evidence from both geology and paleontology that water in the liquid state was present on earth three or even four billion years ago. The earliest known cellular fossils are 3.5 eons old and include blue-green algae, which require water as a living environment. Indeed, the very presence of life itself is difficult to explain without the presence of water.

Geologic evidence for liquid water on the early earth includes the pillow lavas (which are formed by lava erupting under water), mud cracks, and ripple marks in rocks 3.2 billion years old.

Since these facts are well established, we must consider what conditions might have existed in the early history of the earth that would have produced a temperate climate, given a lower intensity of radiation from a younger and weaker sun. One possibility is that the atmosphere of the early earth contained gases which absorbed and retained the heat of solar energy better than the gases that now constitute our atmosphere. Astronomer Carl Sagan has suggested that the presence of small amounts of ammonia and methane would have provided appreciable absorption and would have effectively closed one of the largest "windows" through which radiation now escapes into space. This "greenhouse effect" would have helped to compensate for the weaker solar energy received by the earth during its early history.

Most earth scientists believe that the initial atmosphere of the earth was composed of methane, ammonia, water vapor, and trace amounts of other gases, a mixture very similar to the atmosphere on Jupiter today. No free oxygen existed at that time, although oxygen was continuously produced by sunlight dissociating water molecules. The hydrogen atoms, being very light, escaped into space, but the oxygen reacted with methane to form carbon dioxide and water vapor. It also reacted with ammonia, forming more water and free nitrogen, which gradually built up in the atmosphere. At the same time volcanoes, hot springs, and fumaroles were expelling steam, carbon dioxide, carbon monoxide, sulfur dioxide, and nitrogen. After all the methane and ammonia had been converted, free oxygen could accumulate as more water molecules were dissociated by sunlight. Later, photosynthesis by living organisms increased the concentration of oxygen in the atmosphere.

According to this scenario there would have been a very strong greenhouse effect to hold in the warmth of sunlight during the early period of earth's history. It would have slowly diminished as the amounts of ammonia and methane were depleted. Fortunately, at the same time the sun was growing stronger. But there may have been a time when the methane and ammonia were totally depleted and the sun's energy was still not producing enough heat to compensate. If this occurred about 800 to 600 million years ago, the result could have been an era of very cold temperatures, an effect sufficient to cause a full ice age. This combination of events might explain one of the earliest great glacial epochs: the Precambrian, which appears to have been so widespread that it affected even the equatorial regions.

~❧

The earth's atmosphere acts as a delicate screen and filter controlling the amount of sunlight absorbed by the planet and the amount of heat energy reflected back into space. The balance is altered not only by the gases that are present but also by the myriad tiny particles that are carried in the air, such as motes of dust, cloud droplets, and volcanic ash. Very large volcanic eruptions send tremendous volumes of steam, sulfur dioxide, and finely divided particles high up into the atmosphere above the region where rain might carry the volcanic debris back to earth. Important eruptions are known to have produced a slight cooling effect on the climate lasting for a year or two.

Benjamin Franklin was one of the earliest scientists to suggest that volcanic eruptions could influence the weather. He observed that a "permanent dry fog" covered North America and Europe during the summer of 1783 following volcanic eruptions in Iceland. The sun's rays were dimmed and Franklin surmised that the reduction in solar energy received on the earth's surface might have been responsible for the severely cold winter of 1783–84.

Following the April 1815 eruption of Tambora in the Dutch West Indies, in 1816 northern Europe and New England experienced a year without summer. Halfway around the world and fifteen months after the event extensive crop failures occurred. New England had frost every night except for two periods slightly longer than three weeks each. Unusually violent storms raged and the average temperatures recorded were a few degrees below normal. In 1884, the year following the eruption of Krakatoa, a sharp freeze occurred in California, as evidenced by studies of tree rings. In 1963, after the eruption of Mount Agung on Bali, observations at Mauna Loa Observatory in Hawaii showed a decrease of about half a percent in the solar radiation reaching the earth's lower atmosphere. An eruption of El Chichón in southeastern Mexico in April 1982 threw an unusually large cloud of volcanic debris into the atmosphere. In less than a month it had circled the globe, scattering sunlight and reducing by perhaps as much as 10 to

20 percent the solar energy reaching the earth's surface in Hawaii, Japan, and tropical regions. The cloud of El Chichón is comparable to that produced by Krakatoa and could cause a similar weather disturbance for a year or two.

Although the influence of volcanic eruptions on climate has been quite convincingly documented, the effects are relatively short-lived. It is hard to believe that such transient variations might have precipitated an ice age. But theoretically it appears that this could have happened. When an accumulation of ice and snow does not melt in the summer, it reflects the sun's energy more efficiently than bare ground. The heat is sent back into space and not absorbed by the earth. Thus a cooling trend is started that feeds on itself and can lead to a progressive decline in the world's temperatures. Increased temperature differences between the equatorial and polar regions cause more frequent storms, more snow falls and stays on the ground longer. Year after year the buildup continues and the ice pack grows. Eventually, when much of the earth's fresh-water supply is locked up in glaciers, drought conditions become severe; then less snow falls in winter and the cycle is reversed.

Some climatologists believe that periods of very frequent volcanic eruptions could have been responsible for the great ice ages, and information from deep-sea cores taken around the world lends some support to this hypothesis. An unusual amount of volcanic ash has been found in layers corresponding to the period of the last glacial epoch. Many more violent eruptions seem to have occurred during that time than in the preceding eons. Cores taken through the ice sheets in Greenland and Antarctica contain layers of volcanic dust with particularly heavy deposits during the peak of the last ice age, 30,000 to 16,000 years ago. The amounts indicate a level of volcanic activity five or ten times as great as that which we are experiencing today.

In spite of this rather impressive evidence there are difficulties with establishing a causal relationship between volcanic action and ice ages. Although increased volcanic eruptions may have accompanied the glacial periods, we cannot be sure that they were the initiating cause. Perhaps the reverse is true: ice sheet buildup may cause increased volcanic eruptions. The change in loading of the earth's crust when tremendous volumes of water are removed from the oceans and stored in icecaps on the land could cause fracturing and increased volcanic activity. Furthermore, there is no adequate theory to explain why volcanic activity should have waxed and waned in a rhythm corresponding with the timing of the glacial epochs.

It has been generally assumed that volcanism was much more prevalent in the very early stages of earth history when the crust was just forming. Water in the liquid state, however, seems to have been present long, long before the first known glacial epoch.

Another possible explanation of ice ages is based on the theory of worldwide crustal movements. As we can see in the world today, the distribution of oceans and landmasses affects the patterns of ocean and air currents. The interiors of large continents experience more severe weather conditions than coastal regions and islands. Oceans lose heat very slowly; they are warmer than land in winter so ice floating in the sea melts more rapidly than ice piled up on land. When a continent occupies a position in the polar regions, glaciers can accumulate there, while the presence of an ocean in these coldest parts of the world tends to moderate the temperatures. Antarctica and the Arctic Ocean are examples of these two extremes today. The amount of glacial buildup in the south polar region is ten times the amount present in the vicinity of the North Pole.

During the time of the fourth great glacial epoch (290 to 270 million B.P.) the continents lay close together, forming a nearly continuous mass, and portions of it lay across the South Pole. This configuration was favorable for the accumulation of glacial ice. However, the earlier and later glacial epochs cannot be so readily explained on this basis (see Figure 3). There is not enough evidence either to support or to refute the explanation based on plate tectonics.

In recent years several astronomical explanations of glaciations have been proposed. If a supernova exploded nearby, it would cause an influx of high-energy radiation that would destroy the earth's ozone layer, and a general cooling might occur.

Another theory takes its cue from the long and rather regular periods of time separating the great ice epochs. In 225 million years the solar system makes a complete circuit of the galaxy. If it passes through a cosmic dust cloud in one region in this galactic orbit, some of the solar radiation would be absorbed or reflected by the dust and less would reach the earth's surface. This theory is attractive because it is the only one that explains the wide spacing of the great ice epochs. But astronomers have searched in vain for the presence of such a cloud.

A third astronomical explanation has attracted a great deal of interest and support in the last few years. It is based on the fact that small periodic variations in the earth's movement around the sun result in changes in the distribution of solar energy on earth. In the 1920s a Yugoslavian scientist, Milutin Milankovitch, drew the attention of the scientific world to these variations, which he believed were sufficient to account for the ice ages. At first the theory aroused very little interest because at that time geologists believed that there had been only four ice ages in the last two million years and the timing of these did not correspond with Milankovitch's theory. But with the discovery of more frequent glaciations the theory has been revived.

The earth's orbit varies slightly from a true circle and this ellipticity changes slowly from year to year, becoming more elliptical and back again in a cycle of 92,000 years. When the orbit is almost perfectly circular, the amount of solar energy intercepted by the earth is constant throughout the year. As the ellipticity increases, there is a slight variation in the radiation that reaches the planet.

Our seasons, however, do not depend upon this very small divergence in distance from the sun. They result from a much larger and more important factor: the tilt of the earth's axis, which causes different portions of the planet's surface to face away from or toward the sun. The warmest weather occurs in the equatorial regions where solar radiation follows the most direct path through the atmosphere. At higher latitudes the solar radiation must pass through greater and greater amounts of atmosphere. It strikes the surface at a glancing angle; so the effective radiation reaching the ground is reduced. These seasonal changes occur in a regular cycle throughout the year and they are more prominent with increasing distance from the equator. Slight changes in the tilt of the earth's axis also occur in a long cyclical pattern, swinging from 22.1 to 24.5 degrees and back again in 40,000 years.

Another perturbation in the earth's movement has been observed for many centuries. Its effects were noted as early as 150 B.C. by the Greek astronomer Hipparchus. At the spring equinox, halfway between winter and summer, the sun is at a definite place in relation to the other stars in the sky and returns to approximately that same position the next year. But not quite the same place. It gets around to the same patch of stars about 20 minutes too late, and the next year there is another lag of 20 minutes. This slow creep around the celestial sphere has been called the precession of the equinoxes. Its progress has been watched through the centuries and is found to amount to about 1½ degrees a century, thus completing a full circle in 21,000 years. The reason for the creep has now been identified as a wobble in the earth's axis due to gravitational interactions of the moon and the sun on the earth's equatorial bulge.

These three separate phenomena, each with its own period, affect in small but significant ways the amount of radiation received at any point on the earth. Odd as it may seem, the times of maximum eccentricity and maximum tilt provide the least favorable conditions for glacial buildup. At these times high latitudes receive more solar radiation during the summer months and ice accumulated from the previous winter is melted. The three periodic variations in the planet's motion set up a complex but predictable rhythm, sometimes reinforcing, sometimes canceling each other. With modern computers their combined effect can be calculated. When a curve showing the theoretical rise and fall of the total amount of radiation received at 65 degrees north latitude during the summer throughout the last 600,000 years was compared with the variations in temperatures calculated from oxygen isotopes and fossil marine species in ocean cores, an impressive correspondence was found. For this reason the Milankovitch model is quite widely accepted today. Predictions based on this theory are not reassuring. They foretell a much colder climate, and extensive Northern Hemisphere glaciation over the next 20,000 years.

While the Milankovitch theory provides a very promising explanation for the rise and fall of glaciations within an ice epoch, it does not explain the larger-scale pattern. The temperature changes set up by deviations in the planet's movement around the sun occur much too frequently to account for the timing of the major ice epochs. And they are too small to trigger a glacial age except when some other factors also reduce the earth's temperature. For example, an abrupt shift of earth mass could alter the angle of tilt just as a spinning gyroscope can be made to precess by giving it a nudge. Perhaps relatively rapid movements of crustal plates caused such imbalances. Other possibilities include the impact of giant meteorites that gave the planet a nudge large enough to change the wobble and the tilt of the axis. Meteorites of considerable size have dug many impact craters on other planets in the solar system as recent space explorations have shown. On earth, scars created by projectiles from space are more difficult to identify because they are rapidly obscured by erosion and covered by vegetation. But other clues to such events may lie embedded in the rocks. Perhaps also they are reflected in the waxing and waning of the great ice epochs.

In spite of this dazzling array of imaginative ideas a completely satisfactory explanation of the great glaciations still eludes earth scientists. In the meantime an appreciation of the myriad ways that ice has altered the world grows with every passing day.

Life forms have made many adaptations to severe winter weather. Deciduous plants drop their leaves and pass through a dormant period, just barely sustaining life until the returning sun brings them back into full bloom. Some plants drop all growth exposed to the weather and store up energy in bulbs beneath the protecting soil. In a similar way animals, too, have evolved methods of hiding from the winter blasts: digging deep burrows, storing up food supplies, and passing through the long siege of cold weather in a dormant state of reduced activity and lowered metabolism. Animals that can travel swiftly and whose food supply is available over a wide area escape the rigors of the worst weather by seasonal migration: herds of caribou, seals, penguins, and great blue whales follow the temperature conditions they prefer, traveling twice a year over hundreds of miles. Birds, the most mobile of all vertebrate species, have dealt with the challenge in a way that inspires the admiration and envy of man.

When the days grow short and flights of geese are silhouetted against a gray November sky, their mournful honking echoes between gathering storm clouds and the dying leaves of the forests. These sights and sounds send a chill into the hearts of those who cannot follow the sun south. They warn of the return of seasonal ice, the onset of the little little ice age, which descends with implacable regularity year after year in the temperate and polar regions of our planet.

Soon the frost flowers will grow on the windowpanes, the wintry blasts will shake the last dead leaves from the trees, and the ground will grow hard beneath our feet. It is tempting to think then that the earth must have been a much more delightful place to inhabit during the long uninterrupted periods when warmth bathed even the middle latitudes all year, when trees remained green, the air and earth soft in every season. It is easy to forget how much we owe to ice.

When a large part of the earth was warm, the land was covered with lush forests; shallow seas and swamps were widespread; and the oceans teemed with aquatic life; but the fascinating variety that we enjoy today did not exist. The air, heavy with moisture, moved sluggishly because there was no large temperature difference between the poles and equator to set up strong air currents. Clouds covered much of the sky, forming from the water vapor in the hot, humid air. Weeks or months may have passed when the sun's face was hidden. There were no clear, crisp days such as we enjoy when the sun shines with golden warmth in an azure sky, and the air is cool and sharp. There was no season like our fall when the forests put on a magnificent display of gold and crimson and copper-colored foliage. Without the ice we would not suffer from the harshness of winter but we would never know the joys of spring.

❧

Spring is the supreme gift of the Ice King. Every year it reenacts for us the drama of life transforming the barren earth. A few weeks — just a little fraction — out of any lifetime can be well spent observing this extraordinary unfolding of life in all its exquisite detail.

This year I watched spring come to a little plot of woodland near my home in Illinois. It was a small, protected glen where the southern sun shone strong through the still-barren branches of beech and hawthorne trees and a bank on the north held back the coldest winds. A tiny rill whispered over shining stones before it sank into a deep pool where the flow was held back by the decaying log of a giant oak tree. Along the edges of this pool the grass was beginning to turn a fresher green. A mat of periwinkle, its old leaves bronzed here and there with winter frost, was putting forth tiny pointed new leaf buds along its stems. A few early midges danced in a shaft of sunshine.

In this sheltered place I sat down on a cushion of moss. Carefully pushing aside the thick mat of oak leaves and pine needles, I uncovered a whole miniature world of resurgent life. In this little space a dozen different plants had pushed firm tips of dark curled leaves through the soil that was still cool to the touch and crumbly in my fingers. I could not yet guess what shape would grow from each of the thrusting tips. The identifying bulbs and roots lay hidden beneath the spongy soil where I could see and feel the stirrings of life as bacteria and earth-worms wove their blind patterns of activity in the darkness.

The days passed and the earth warmed. I returned each morning to watch the individual shape of each plant unfurl. First came the dis-tinctive heart-shaped leaves of the violet and then its sweet-faced lavender blossoms. The windflower shook out its feathery foliage in the cool breeze. The dark red tip of the bloodroot pushed upward through the moist leafmold, taut and purposeful as a phallus bearing its carefully shielded seed of life. It rose high above the tangled mat of vegetation. Then gradually the deeply lobed leaf unfolded, resigning its precious charge, a single bud that slowly swelled into a waxy flower with a golden center.

On slender, tremulous stems the twin-flower released pairs of nodding pink blossoms high above its satin leaves and filled the soft woodland air with a delicious fragrance.

The dark flags of large-flowered trillium slowly unfurled to reveal the precise mathematical perfection of its form. Three symmetrical leaves opened out in unison, forming a shallow cup that bore a tiny green bud in its center. The bud grew and the green calyx split into three pointed sepals that curled back, releasing the blossom. And the blossom unfolded its triangular petals in unison, culminating in the perfect three-lobed flower. Three upon three upon three, an equation written in time.

Fiddles of the cinnamon fern pressed upward toward the sunlight, raising their tightly packed whorls on delicate green stalks. These intricately involuted scrolls looked like the shapes of the great spiral galaxies scaled down to the size of my thumb. I watched the coils of organic matter expand and tiny tentacles reach out, groping toward the sunshine. If I were as big to a galaxy as I am to a fern, I thought, I could watch it unfurl and reach pale fingers out into space to catch the cosmic-shine.

Overhead, the birch trees began to put forth a fretwork of silvery buds that seemed to hold the light, giving it back slowly as though they each had a radiant source of their own. Against the deep shadows of the woods these buds of light looked innumerable and yet as distinct as star clusters in the Milky Way. The flowering of spring here in this little glen was a tiny universe compacted in space and time.

All too soon the season ripened into summer. The petals fell from the wood anemone and the foam flower. The fringed maple flowers were replaced by dark green leaves; the delicate tender green of spring gave way to a heavy canopy of leaves. Now the rays of light could not penetrate the dark shadows, and the ground was wet and soggy under-foot. No light breezes stirred the humid air. So might the world have seemed when jungles dominated the earth and no glaciers floated in the polar seas.

❧

Ice gave us the fragile beauty of spring, the returning robin, the snow goose, and the crocus, and perhaps something more. Mankind grew up during the Pleistocene ice epoch, and the rigor of the glacial climate may have accelerated his evolution. Although the theory is speculative and can probably never be proved, it has been suggested that the challenge of a deteriorating environment was a factor leading to the rapid development of the brain. As the birds invented seasonal migrations, man developed unique human responses. In order to survive the severe winters he learned to plan ahead, to join in cooperation with others of his kind, to build shelters, to cover his body with animal skins, and he discovered the use of fire. In a world of eternal summertime where the living was easy the mysterious potential of mind might have remained dormant. The symphonies of Beethoven, the poetry of Keats, the space probes to Saturn, and messages to the stars — seeds of these wonders and others still unknown have been dropped one by one from the retreating glaciers, released from the shining splendor of ice.

The Edge of the Sea

Only, from the long line of spray
Where the sea meets the moon-blanched land
Listen! You hear the grating roar
Of pebbles which the waves draw back, and fling
At their return, up the high strand,
Begin, and cease, and then again begin ...

— MATTHEW ARNOLD, "Dover Beach"

On a windy August day I stood on the bluff of Pointe du Raz, where the coast of Brittany juts a rocky elbow out into the Atlantic, and watched the waves pour in from a stormy sea. Stirred and driven by the wind, the ocean was a tumbling mass of white spindrift and deep blue water. It broke against the dark rocks far beneath my feet, and sent plumes of salt spray against the jagged cliffs, diffusing the air with mist. The wind in my face, the roar of the waves in my ears, the sight of the turbulent water filling my view from serrated horizon to the foaming shore — what joy there is in feeling the rhythm, the tempo, and beat of the sea!

More than any other place on earth, here on the ocean's tumultuous threshold is focused that mysterious force — the restlessness of the universe without which nothing would *be*. Without it the planets would fall in on the sun. The sun and the stars of the Milky Way would collapse on its glowing center. The galaxies would stop fleeing from each other at unimaginable speeds. The cosmos would fold up like a morning glory closing its petals. Time itself would come to an end when all movement stopped.

Although we seldom wonder why, the sight, the sound, the feel of motion is a lovely thing. The sailplane pilot knows it as he soars on the wind, the skier as he plummets through the icy air on the mountainside, the dancer as she floats on waves of sound. And everyone is exhilarated by the ceaseless pounding of the surf, the sweep of vagrant breezes, the play of light and shadow on the ephemeral, ever-changing margins of the sea.

Along the dark shaly strands where pebbles roll endlessly to and fro, on a thousand wild rough shores where rocks and sea match their strength, even on the deceptively quiet beaches where row upon row of delicate white ruffles edge the skirt of the sea, action and change are always taking place. The interface between land and water varies from minute to minute as the waves roll in, break, and slip back to calm depths. It moves from hour to hour with the rising and falling tides. Even from eon to eon dynamic activity has been concentrated along the shores.

～⁓

The vitality of the ocean's edge is expressed in many ways. It is a favorite habitat for living things. Seaweeds and barnacles, limpets and periwinkles cling throughout their whole adult lives to rocks that are battered by surf and tide. Crabs and sea turtles live part on land and part in the sea, scuttling back and forth across the beach where they lay their eggs and dig protecting burrows beneath the sand. Mangrove trees extend long prop roots down into the mud and sand, building little islands and redesigning the shape of the shore.

A rich assortment of aquatic organisms inhabits shallow coastal waters, which the sun warms and permeates with radiant energy. Here the microscopic plant forms bloom in bewildering profusion. This phytoplankton, composed of tiny vegetables of the sea, makes the energy of sunlight and the mineral wealth of the ocean water available to the diverse animals that feed upon it.

The word *plankton* comes from the Greek, meaning "wandering." Like plants on land these tiny organisms cannot choose their habitat. They float and bloom and die wherever the winds, tides, and currents take them. Great drifts accumulate in the shallow water above the shelves, in restricted bays and inlets along the coasts. There they provide a plentiful base for the aquatic food chain: the microscopic zooplankton, hordes of tiny carnivores, crustaceans, fish of many sizes. Ninety percent of all the seafood we eat or use as animal feed and fertilizer is harvested from waters overlying the continental shelves.

A remarkable "alchemy" takes place wherever oceans receive river waters because fresh and sea water are very different in composition. Many minerals are carried dissolved in water, so finely divided that they are invisible there. The taste sometimes reveals their presence — the metallic flavor of iron in well water, the bitter tang of salt in the sea. Where the two kinds of water meet at the river mouths and deltas, chemical reactions take place. Suddenly crystals materialize and collect, building up layer after layer of solid formations.

The tides and changing ocean levels alternately fill and drain swamps and tidal pools, exposing them to the energy of sunlight, then covering and sealing them off from the air. This ceaseless activity, working patiently over millions of years, has created and concentrated stores of valuable minerals in many places that at some time lay on the edge of the ocean.

~❧~

Salt was the first mineral gift from the sea to be discovered and used by man. Deposits of this important compound accumulate in arid climates where shallow bays are inundated by rising tides. Frequently the bay has a sandbar or sill along its entrance channel; so the water does not all drain out again when the tide retreats. Some of it is trapped in the basin and is heated by the sun. Evaporation occurs, removing water molecules, leaving salt and other compounds behind. If evaporation exceeds rainfall, this process gradually concentrates the solution. When it becomes approximately ten times as strong as normal sea water, the salt precipitates out, making beds of white crystals along the shores. Under special conditions of very high evaporation and minimal rainfall, salt deposits can be laid down within continental landmasses on the perimeters of drying lakes. But the majority of salt deposits are formed on the shores of the sea.

Salt is an essential chemical for living cells. It facilitates the transport of proteins and fats by body fluids and allows them to function properly within the cell. In higher organisms salt is especially important for communication among nerve cells in the brain. Thus the chemistry of the ocean water followed the vertebrates onto dry land

and has driven them throughout countless millennia to seek out those places on the earth's surface where stores of salt crystals have been created by the air, the sun, and the sea.

In the most primitive human societies salt was obtained secondhand from animal flesh but with the development of agrarian societies salt had to be added to diets that consisted largely of grains and fruit. Later, the discovery that meat could be preserved by salting created an even greater demand for this mineral. So salt became very precious, its value increasing with distance from its sources. It was so highly prized that it was used as a form of currency and a source of tax revenue, customs that still survive in some places today. Marco Polo observed and reported these usages in thirteenth-century China:

> In this country [Kain-du, a western province] there are salt-springs from which they manufacture salt by boiling it in small pans. When the water has boiled for an hour, it becomes a kind of paste, which is formed into cakes of the value of twopence each. These, which are flat on the lower, and convex on the upper side, are placed upon hot tiles, near a fire, in order to dry and harden. On this latter species of money the stamp of the Great Khan is impressed, and it cannot be prepared by any other than his own officers. Eighty of the cakes, in proportion as they find the natives less civilized, are carried by the traders amongst the inhabitants of the mountains. . . . Their profits are considerable, because these country people consume the salt with their food, and regard it as an indispensable necessary; whereas the inhabitants of the cities use for the same purpose only the broken fragments of the cakes, putting the whole cakes into circulation as money. . . .
>
> Upon salt . . . [the Great Khan] levies a yearly due of eighty tomans of gold . . . amounting to six million four hundred thousand ducats. This vast produce is occasioned by the distance of the province to the sea, and the number of salt lakes or marshes, in which, during the heat of summer, the water becomes crystallized, and from whence a quantity of salt is taken, sufficient for the supply of five of the other divisions of the province.

Throughout the ancient world, trade routes were set up to transport salt. The commerce between the Aegean and the coasts of southern Russia was largely dependent on the salt pans at the mouth of the Dnieper River. The salt of Palmyra on the edge of the Syrian desert was traded between the Syrian ports and the Persian Gulf. Salt mines in northern India were worked before the time of Alexander, in Austria and Syria in Roman times.

Even today one can watch camel caravans winding slowly along the blazing desert of the Danakil Depression in Ethiopia, laden with blocks of salt for the markets of Makale, where it is used as a form of local currency just as it was in the days of Marco Polo. The Danakil lies near the Red Sea in the Afar Triangle, that remarkable piece of earth surface which is rifting apart as a new ocean is being formed. The center of the Triangle is a basin lying below sea level — in some places almost 400 feet below. It is extremely arid: average yearly rainfall is less than an inch a year. Deposits of salt and other evaporites almost 3,000 feet thick have formed there in the last two million years. Their enormous depth suggests that this land must have had intermittent connections with the Red Sea.

Evaporites can be seen forming now as the lakes and ponds continue to dry up. Shallow bodies of water like Lake Asal are surrounded by glittering fields of salt crystals and gypsum (a compound of sulfur). Every stone or little protrusion is crowned with an array of iridescent shapes, delicate as spun glass. A bottle left by a careless tourist just a few days before I visited the site was completely encrusted with long, shining needles of gypsum that were translucent, pale ivory-colored, and so fragile that they shattered at a touch. The brittle sheet of crystals encircling the level shores of the lake, like thin ice, broke through as I walked across it and I descended suddenly into black mud as soft and treacherous as quicksand.

The water of Lake Asal is very warm, heated by the sun and the many hot springs that feed it. But the rate of evaporation is so high that no water flows out. The chemicals that are carried in the water become more and more concentrated there. Gypsum, a normal component of sea water, is precipitated by the same evaporative process that produces salt deposits but at slightly lower concentrations. It begins to crystallize out when the solution is about five times as strong as average sea water.

Gypsum is not essential to health like the salt with which it lies so closely bedded, but it has proved its worth in many less important ways. In ancient Egypt it was used to make the plaster that covered the walls of many tombs and temples, forming the smooth surface on which many beautiful paintings were executed. Now gypsum is employed widely in the building industry to make wallboard, plaster, and cement. In pulverized form it is utilized as a soil conditioner and fertilizer.

❧

Another type of mineral deposit, which is very valuable as fertilizer, is often found associated with evaporites. These rocks — called phosphates — contain compounds of the element phosphorus. In its pure form phosphorus is a remarkable element that glows in the dark and catches fire when exposed to air. One of the building blocks of every living molecule, it does not occur free in nature, but its compounds are present in many rocks and minerals, comprising about one tenth of one percent of the lithosphere. Since it is taken up and concentrated by vegetation, agricultural soils become depleted of phosphorus and it must be replaced to maintain fertility. So deposits of phosphate rock are important natural resources. These rocks are found in places where marine life has been abundant.

Sediments on many continental shelves are rich in phosphates because those waters are a favorite habitat of aquatic life and organic matter accumulates on the bottom of the sea margins. Conditions for the deposition of phosphates are especially favorable where cold, nutrient-rich bottom waters flow up from depths near the shore. These upwelling currents occur most frequently on the western continental coastlines where ocean circulation caused by the rotation of the earth drives the surface water away from the shore and allows the deeper waters to rise, thus providing an abundance of essential nutrients that support large communities of marine organisms. Off the coast of Peru, for example, an immense population of anchovies has supplied one of the largest fishing industries in the world. In some years the yield from this one area alone has exceeded 15 percent of the total harvest of marine animals.

Although the chemical reactions that cause precipitation of phosphates are not well understood, they appear to occur most frequently in warm, arid locations on shallow shelves where the sea water is quite saline.

❧

Many other treasures of great commercial value have been stored up in restricted, shallow-marine environments. Iron salts eroded from the continental rocks are carried in river water in a finely divided suspension. When they reach the ocean, oxidation occurs, producing compounds that are insoluble in water. Many of the iron deposits of the earth — at least, those younger than 1.7 billion years — were deposited in shallow estuaries on the ocean margins.

There are two types of iron deposits on earth and they differ strikingly in appearance. The very ancient iron formations occur in dark layers alternating with light gray or pink layers of chert. The origin of these banded-iron deposits is one of the great unsolved puzzles of geology. It seems to have involved an alien chemical environment — probably a different atmosphere and sea-water composition from what is present in the world today. These oldest deposits are very extensive, covering hundreds of square miles. For example, a belt of banded-iron ores stretches from Labrador through Quebec, Ontario, Wisconsin, Michigan, and Minnesota. The banded-iron formations represent by far the richest iron deposits in the world, a legacy from that shadowy time before multicellular organisms began to leave a detailed history of the earth in the rocks.

But the iron deposits that have formed in the last 1.7 billion years and that are believed to have precipitated on the edge of ancient seas have also played an extremely important role in human history. The discovery of Middle Jurassic iron ores (150 million years old) in England ushered in the industrial age. Ores of similar antiquity in Alsace-Lorraine were fought over by France and Germany for many centuries; possession of these mines was one of the causes of the Second World War.

≈

The wars of the immediate future will probably be fought over possession of still another kind of resource laid down in that transition zone of land and sea. Shallow, sunlit waters of the continental shelves provide ideal conditions for the creation and accumulation of oil. These stores of "distilled sunlight" were created by living plants converting the radiation of the sun into carbohydrates, which were then sealed off and preserved in the crust of the earth by the moving waters and drifting sediments of the land.

Profuse blooms of algae and plankton occur on the margins of the sea in amounts that often exceed the quantities consumed by larger organisms. As we saw in the case of phosphate deposits, the densest concentrations of marine life occur in those places where upwelling currents bathe the shelf in cold water bearing essential nutrients in suspension. Living and dying by the billions and billions, the bodies of these minute marine organisms carpet the bottom of the shallow basins. As the deposit of organic sediment increases, it exceeds the capacity of the oxygen dissolved in sea water to decompose it. In some especially favorable locations in protected bays and estuaries oxygen is rapidly depleted because little agitation of the waters occurs.

Wherever rivers carry sediment eroded from the land and deposit it on the shelf, the organic remains of marine life are buried and sealed off, further impeding decay. Only anaerobic bacteria can survive these conditions. As the weight of the sediment increases, often to depths of several miles, the pressure, heat, and anaerobic bacteria working slowly over many eons turn the organic debris into oil and gas.

Source rocks containing filaments of petroleum in their pores are further squeezed and compacted; the oil oozes out into adjacent rocks. Under the most fortunate circumstances it may find a resting place in porous formations like sandstone or limestone — perhaps an old coral reef. And if a cap of impervious rock like shale or salt is present, preventing further diffusion, the oil accumulates in these reservoirs.

Considering all the special conditions that must be met in order to create an oil deposit, it is surprising that so much of this valuable resource has been discovered and many more reserves are believed to exist on the continental shelves. Although about 80 percent of the oil wells in production today are on land, most of the areas where they are located lay on continental shelf sometime during the past 200 million years.

The age of oil deposits encompasses a relatively limited period of geologic time. Their formation requires time for a deep layer of sediments to be laid down over the organic remains and to subject them to heat and pressure conditions sufficient to turn them into oil — a process that usually takes at least a million years. Once formed, the reservoirs that hold the precious fluids do not remain intact forever. When they are fractured, the oil diffuses out and is lost. Reservoir rocks are usually younger than 200 million years.

But wherever oil is found, though it be beneath deserts or on barren plateaus, was once the bottom of an ocean margin. As water levels rose and fell in response to the ice ages and movements of the earth's crust, the location and shape of the shelves have changed. About 100 million years ago the Mideast oil fields around the Persian Gulf lay beneath a shallow sea; limestones and coral reefs have left their signatures there. Along the Gulf of Mexico sediments are now as much as eight miles thick, but about 100 million years ago waters extended far inland, covering much of the central portion of the North American continent. The oil that was formed on what is now dry land during these periods of high water levels were the most accessible reservoirs and, therefore, the first to be discovered.

As early as 3,000 years before the Christian era oil deposits attracted the attention of man. In Mesopotamia on the banks of the Tigris and Euphrates rivers north of the Persian Gulf a vast reserve of oil lay just beneath the surface. Seepages of a dark semisolid substance — an asphaltic bitumen — were found at Hit on the Euphrates and several other locations throughout this area. The Sumerians found that this gummy material was valuable in many ways. It was used for waterproofing ships and wicker baskets, as a bedding compound for mosaic and inlay work. Mixed with sand and fibrous materials, it served as a mastic in architectural and road construction. Softened with olive oil, it was rubbed on sores, open wounds, and rheumatic joints. Administered with beer, it was taken internally and offered as a cure for a wide range of ailments. The smoke from burning bitumen was believed to frighten away evil spirits.

The techniques associated with these uses of bitumen, developed over almost 3,000 years, were lost with the decline of the Sumerian civilization because knowledge of them was not transmitted through the Greeks and Romans to the Western world. Other products more readily available in Europe — tars and pitches from wood, lime, and volcanic rocks — were used by the Roman engineers for building purposes. Bitumen was relegated to the realms of magic and medicine.

Belief in the therapeutic virtue of petroleum was very widespread, persisting up until recent times. Wherever seepages of bitumen and oily deposits were discovered, people used them as ointments for aching joints, rheumatism, toothaches, sprains, and burns. These treatments were reported to give great relief. The medicinal uses led eventually to the drilling of the first oil wells in the United States and to the birth of the modern petroleum industry.

In the early decades of the nineteenth century oil flowing from springs in western Pennsylvania into a tributary of the Allegheny River was collected, bottled, and sold as "Seneca Oil" (this name was probably derived from the medicinal use of oil by the Seneca Indians). Farther south in the Virginias oil was also being found at that time but in circumstances where it was viewed as a troublesome contaminant and was regularly discarded. An important salt-mining industry had been active there for many years. Salt licks and springs that yielded brine served as the earliest sources of this important commodity for the pioneers who were opening up the country west of the Alleghenies. But boiling the saline spring water produced salt of inferior quantity and quality, so tools and drilling techniques were soon invented for extracting brines from greater depths. By the 1830s wells of 1,000 feet or more were satisfying the large demand for salt to be used in preserving meat and fish, or curing hides for shoes and harnesses.

In many of these salt mines a thick brown fluid appeared along with the brine. Since this oily substance threatened to destroy the purity of the salt, it was treated as a waste product and allowed to flow over the top of the well into the nearby streams and rivers. At one salt mine in Kentucky thousands of barrels of oil gushed out and flowed into the Cumberland River. Two miles downstream a small boy set the oil on fire. The conflagration raged along a fifty-mile stretch of the river, destroying the forests along the shores and leaving scars that lasted for many years. The "ruined" salt mine was taken over by local promoters who formed the American Medicinal Oil Company in the late 1830s. Several hundred bottles of "American Oil" were sold in the United States and Europe.

At about this same time in Tarentum, Pennsylvania, a man named Samuel Kier drilled for salt on his farm and struck saline water of good concentration. When he found that it was contaminated with an oily substance, Kier drained off the brown fluid and dumped it into the nearby Pennsylvania Canal.

One day Kier's wife became ill and the doctor prescribed "American Oil." Kier observed that this expensive medicine looked and smelled just like the contaminant he had removed from his salt well. A little investigation proved this hunch to be correct and in 1849 Kier opened a business in Pittsburgh, selling the oil bottled from his own salt mine and a mine leased from his neighbor. He called the product "Rock

Oil," and promoted it with outstanding success. Handbills simulating banknotes proclaimed its therapeutic virtues: three doses taken daily could make the blind see and the crippled walk; when used as a liniment it could cure rheumatism, gout, and neuralgia. In the center of each banknote was a drawing of the drilling rig at Tarentum. One of these handbills, displayed in the window of an apothecary shop in New York, caught the attention of George H. Bissell. It was the summer of 1856 and rumors had been circulating around the city concerning the possibility of using oil made from coal as a fuel for lamps. Perhaps, Bissell thought, the oil from mines like Kier's would be "a good illuminator." This inspiration led three years later to the drilling of the first commercial oil well in the United States.

In a little more than one century the industrial world's insatiable thirst for oil grew from the thirty barrels a day produced by this first well to sixty-two million barrels a day.

Oil exploration is now turning to the continental shelves as the most promising source, especially those shelves along the partially landlocked seas like the Baltic, the Gulf of Mexico, the Arctic Ocean, and the Bering Sea because the depth of sediments in these locations is favorable for producing oil reserves. Geologists believe that particular attention should be given to those areas where upwelling currents may have existed in the past 200 million years. The pattern of changing sea levels during this time is a matter of intense interest to the oil companies. They jealously guard the information on this subject discovered by their scientists.

It may seem paradoxical that some of the richest oil reserves in the world formed in situations that we might consider to be inimical to life: in the Arctic regions, in arid deserts on the edge of drying seas where temperatures sometimes reach 130° F. But we must remember that conditions which seem unfavorable to us may provide an ideal habitat for some life forms, especially the most primitive ones. Plankton and other microorganisms grow profusely in very cold waters enriched by upwelling currents. Many places in the Arctic waters today support a great profusion of aquatic life. On the other hand, some types of algae thrive in saline conditions under very high temperatures and they multiply prodigiously because there is no competition from other organisms. Near Abu Dhabi on the Persian Gulf extensive mats of algae can be seen growing on evaporite deposits. Reacting with the gypsum, they produce foul-smelling sulfur gases, and veins of organic matter extend down through the evaporite layers. This may be an oil reserve in the process of formation.

~&

While petroleum is derived mainly from marine plankton, coal is formed from terrestrial plants that once grew on the land bordering the sea. And many of the important coal deposits of the world are much more ancient than oil. They were laid down in the Carboniferous Period, between 345 and 280 million years ago. At that time North America lay diagonally across the equator and widespread inland seas frequently submerged much of the continent. They formed, drained, and reformed many times during this period. The great extent and shallow depth of these "epeiric" seas damped out the variations caused by lunar tides. Seas of this kind are not known in the world today; the only close modern analogy is the Bermuda Bank. However, they were a prominent feature of the ancient earth.

At the beginning of the Carboniferous Period epeiric seas covered most of North America and large parts of Europe, Asia, and North Africa, as evidenced by the limestone deposits rich in aquatic remains that have been found across broad areas of the continents. When the seas slowly retreated, deposits of mud were laid down near the shores on river deltas and these hardened later into shale. Gradually the continental rocks emerged from the shallowing sea. Rain and wind eroded their surfaces and distributed the fragments. After long periods of deposition and compaction these eventually became layers of sandstone.

A striking repetitive pattern: limestone-shale-sandstone is found throughout the Carboniferous rocks, repeating every two or three million years. As many as fifty cycles stacked one upon the other have been traced across the continents, recording the rapid fluctuation of sea level. The reasons for such frequent rise and fall of water levels are not clear, but they may have been related to the advance and retreat of continental glaciers in the ice epoch that is known to have affected the regions near the South Pole at that time (see page 141).

The coastal swamps on the edge of the epeiric seas were covered with junglelike vegetation. In the luxuriant forests many now-extinct types of ferns flourished and large scaly-barked lycopsid trees. The fossils of these trees have no growth rings and, therefore, we surmise that they grew in a tropical environment without seasonal changes. Remains of distinctive species, such as a certain type of lycopsid tree (the genus *Lepidodendron*) that thrived for only a short period of geologic time and then became extinct, are very useful in determining the time when formations were laid down. They are called "index fossils" by paleontologists.

At times of rising sea level the vegetation that died in the tropical forests fell into shallow water where it was protected from decay and from the aerobic bacteria that break down the energy-packed carbon compounds, releasing carbon dioxide to the atmosphere. The hydrocarbon compounds were slowly buried deeper and compacted by overlying material. The dead vegetation became peat and finally coal — the grade of the coal depending on the degree of compaction. Many seams of coal are embedded in the shale between the sandstone and limestone strata in the "cyclothem" formations that are so characteristic of this period. The fossils dicovered in the various layers and the patterns in which they were laid down tell fascinating stories about the changing relationships of land and sea 300 million years ago.

～ε

Near the small town of Mecca in Indiana a stream has cut a channel down through ancient layers of rock and has exposed several cyclothems that contain seams of coal covered by dark shales. The seams are not valuable enough to be commercially mined at the present time, but the sites are fine places to study the geology of these formations and reconstruct the past history of the region.

In the late Carboniferous this part of Indiana lay on the eastern edge of the Illinois Basin, a large shallow depression drained by an extensive river system. Transgressions of the sea occurred intermittently and inundated the basin. The layers of rock in the Mecca formation show the typical cyclothem: terrigenous deposits of sandstone containing fossils of land plants, then layers of coal and shale, then limestone with remains of marine organisms.

The most interesting portions of this formation are the layers of black shale, a rock characteristic of mud laid down in extremely quiet waters where no circulation brings oxygen to the bottom sediments. The black shale at Mecca contains fossils of snail burrows and small mollusks. In this layer also have been found the remains of many large marine organisms: the fossilized skeletons of hundreds of sharks, some as long as thirteen feet, and shells of giant coiled nautiloids having diameters up to three feet. All the skeletons show marks of body-crushing bites, and the gastric fossil remains suggest that the sharks had been eating shark meat. Farther west, where the Illinois Basin was deeper, the gastric remains of sharks show that they had been feeding on invertebrates.

Rainer Zangerl and Eugene S. Richardson, Jr., the paleontologists who made this astonishing find, concluded that an unusual combination of circumstances occurred here on the edge of this epeiric sea. It must have been deep enough at times to accommodate large aquatic species like sharks and nautiloids. Then when the waters began to retreat, many of the creatures were trapped in lagoons and bayous on the shallow eastern edge of the basin.

We can reconstruct in imagination the dramatic events that may have taken place on the margin of this inland sea. As it slowly shrank in depth and width, the number of organisms that it could support also diminished. Large animals like the sharks and nautiloids may have made daily forays into the coastal bays where food was most plentiful. One day they found the passage back to deeper water blocked, perhaps by falling trees or sand driven in by a storm. Floating vegetation, probably algae, covered much of the water surface, and this, piled up by the wind, may have helped to obstruct the channels. Large numbers of sharks were trapped in small lagoons and competed with one another for the rapidly diminishing supply of plankton and crustaceans. When the scarcity of food became desperate, a few of the boldest sharks turned to cannibalism to quiet the pangs of hunger. Blood stained the water; the other sharks responded with primitive frenzy that spread like wildfire. Soon the still surface of the lagoon was whipped into foam by a hundred lashing tails.

Along the banks in the lush vegetation of the forest, reptiles — the first all-land vertebrates — witnessed this savage scene, lurking in the shadows and looking out impassively with unblinking eyes while the sharks destroyed each other in a terrible orgy that few survived.

Slowly the waters quieted; the algae spread out again across the surface of the lagoon as torn bodies sank and settled into the dark muck on the sea bottom. As the years passed, the mud continued to accumulate and was gradually compacted, sealing the skeletons in a tomb of black shale.

The seams of coal that lie below the sharks' tomb are not mined today but as fossil energy supplies become very scarce, men — descendants of those same reptiles who watched the sharks destroy each other from the safety of the forest — may fight over the last crumbs of stored energy and turn on each other in an orgy of hatred that would have *no* victors.

In many places around the planet but most especially at the edge of the sea, sediments of mud and silt and sand are slowly being laid down now, entombing the flotsam of our time: oyster shells and pop-tops from beer cans, plastic Coke bottles and the delicate bones of little sandpipers that chase the waves along the shore. As layer after layer is compacted here, what story will harden into stone? Will there be bones bearing the marks of a terrible conflict? Or traces of radioactive elements gradually disintegrating with time? Will the skeletons of *Homo sapiens* become the index fossils of the Pleistocene?

Somewhere hidden in the shadows, watching the world of man with wise and gentle eyes, there may be a creature who carries in its genes promises for the future — potential for intelligence and tolerance exceeding those of man. So its descendants may be able to meet the challenge of a severely altered environment, to live in harmony with their fellow creatures, to use the resources of the planet wisely. And they will inherit the earth.

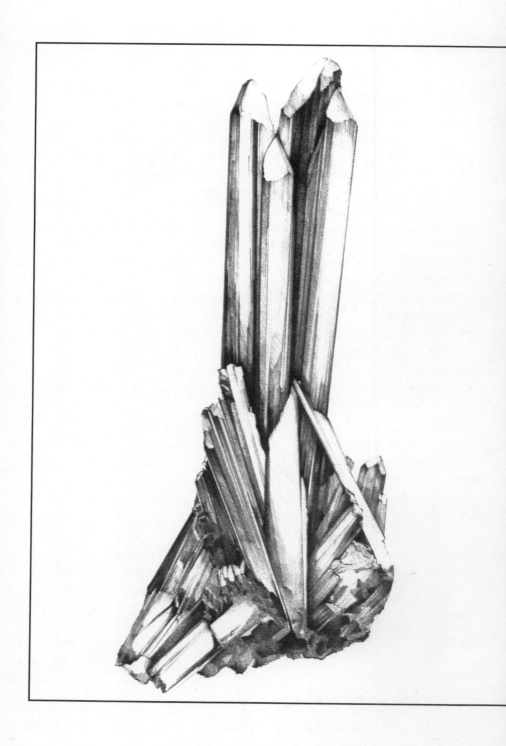

9

Secrets Sealed in a Stone

> Everything is in a state of metamorphosis. Thou thyself art in
> everlasting change . . . so is the whole universe.
>
> — MARCUS AURELIUS, *Meditations*

Fresh snow had dusted the peaks of the Atlas Mountains on the late
November day when I headed north across the range on the highway
from Ouarzazate to Marrakech. The southern slopes rose steeply in a
great wall of tumbled rock and dark green vegetation crowned with
white. At the base of this wall the pale Sahara sands were dotted with
patches of dark pebbles, like the dappled coat of a fawn.

I soon left the desert behind me and the road started to climb steeply.
The sun, concealed from my view behind dark threatening clouds,
threw a shaft of brilliant light deep into a mountain valley where
children were watching sheep on a green hillside and a cluster of
flat-roofed red mud houses was illuminated as on a stage.

When I rounded one of the sharp bends, a half-grown boy ran out
onto the road in front of my car. He held up something violet-colored
that sparkled in the sunlight. I braked to a stop and he approached the
car eagerly, black eyes shining beneath the hood of his long striped
burnoose. He thrust his arm into the car to show me the treasure that
he held in his hand — two halves of a geode broken open to reveal
the beautiful amethyst crystals that filled the cavity with lavender
light and shadow.

I studied the rock carefully because I had been warned that the inhabitants of these mountain villages sometimes dye clear quartz crystals in order to make them more attractive to the unwary traveler. But the violet hues of this geode were distributed in a natural way. The tone was deepest in the center of each crystal, grading smoothly to paler shades on the surface. After some negotiations, we agreed on a price and the geode traveled with me up the many switchbacks on the steep mountain slopes, over the nearly snowbound Tizi N'Tichka Pass to Marrakech, and then to Chicago, where it now sits upon my writing desk.

When I turn the geode in the light, the gleaming facets send rays deep into the purple shadows of the farthest recesses. Long hexagonal crystals with little prismatic faces all point toward the center, directing the rays of light inward like a star turned inside out.

Large crystals are often objects of great beauty and they seem to be a favorite mode of expression in nature: the sapphire, the snowflake, the frost flower. There is artistry in the creation of a crystal and something more profound but only partially revealed. As Loren Eiseley expressed it, matter has arrayed itself in form.

When the two pieces of the geode are fitted together they make a brown, rather undistinguished-looking rock. On close inspection I can see that its gnarled exterior is composed largely of small, irregular quartz crystals, the same ingredient that comprises more than 99 percent of the brilliant amethysts inside. When the geode is closed, it is hard to imagine that it contains anything remarkable. The little Berber boys who roam the shaly slopes of the Atlas Mountains must have learned from long experience how to distinguish between an ordinary rock and one that holds a nest of gems inside.

❧

Inside — this is one of the secrets. Some of the most marvelous creations of nature are found in inside places: a robin's egg, or a hazelnut, or a sunflower seed. All of these build protected inside spaces for their precious contents as they grow. In the case of the geode, however, the space was created first and then the crystals formed.

Cavities are frequently formed by steam bubbles when volcanic rock solidifies from molten lava. Later, water seeps through the outside shell, carrying a solution of silica, a simple compound of silicon and oxygen — elements which together make up three quarters (by weight) of the earth's crust. The enclosed place is shielded from wind and weather. Here nature is able to take its time. Working with infinite pains, it builds an elaborate construction in space, using just two kinds of atoms. One atom of silicon is nested between four atoms of oxygen and this unit is repeated at least a trillion times, spaced side by side with exquisite precision to create the wonderful symmetry of the quartz crystal. A whimsical addition makes the crystal come alive with color and variety. A trace of manganese or iron added to quartz produces rose-colored quartz or yellow citrine or the beautiful violet hue of amethyst.

Free space and uninterrupted time are two essential conditions for the execution of these works of art. But there are not many places in the world where such virgin pieces of space-time are available. Rushed by rapidly falling temperatures, confined to small, cramped places, nature routinely does the best it can with the situation at hand. Most of the crystals it turns out are small and irregular. Quartz crystals like this make up most of the nondescript pebbles and rocks. In fact, a large part of the earth's crust is composed of these poor workaday cousins of the amethyst.

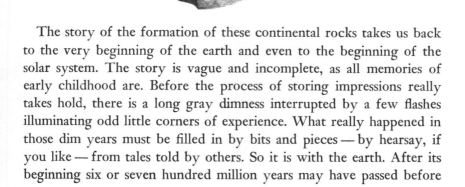

The story of the formation of these continental rocks takes us back to the very beginning of the earth and even to the beginning of the solar system. The story is vague and incomplete, as all memories of early childhood are. Before the process of storing impressions really takes hold, there is a long gray dimness interrupted by a few flashes illuminating odd little corners of experience. What really happened in those dim years must be filled in by bits and pieces — by hearsay, if you like — from tales told by others. So it is with the earth. After its beginning six or seven hundred million years may have passed before the first rocks formed and began to store memories of that long-ago time. What went before must be reconstructed from information brought to earth in meteorites that solidified into rock as early as 4.6 billion years ago.

That date has been taken as the birthdate of the solar system when the sun and all its satellites condensed out of a cosmic cloud. The cloud was composed of hydrogen, the very lightest element in the universe, and a scattering of stardust. All the atoms heavier than hydrogen had been fired in the fierce heat of an earlier generation of stars. These stars — like our sun but much larger — had increased in luminosity until finally they exploded in flashes of energy equal in brilliance to the light of an entire galaxy. The atoms created in the heart of these "supernovae" were blown far away into space, joining the vast clouds of hydrogen that were wandering around the universe.

As time passed, the random motion in the cloud caused local increases in the concentration of atoms; eddies formed and in their centers matter began to draw together under the force of gravity. In the whirling nebula that gave rise to the solar system, much of the material flowed inward and was concentrated in the sun — a mass so large that it retained even the lightest element, hydrogen, within its gravitational field. The portion of the cloud that remained in orbit was very hot, its temperature decreasing with distance from the sun. Gradually small concentrations of matter formed in this cloud and planets began to take shape by accretion. The composition of each planet depended upon the elements and compounds which could condense at that location in the cloud as it cooled. The large outer planets and many of the meteoroids were born in the farther regions of the cloud where temperatures were moderate and even volatile elements were retained. The inner planets — Mercury, Venus, Earth, and Mars — formed close to the sun in a region where the intense heat drove the volatile elements away, leaving the denser particles of stardust to make these small, compact worlds. Mercury and Venus took shape in a very hot portion of the cloud, and water was not able to condense out at those high temperatures. But the earth happened to form in an especially favored location where water could condense from the solar nebula and later could exist on the planet's surface in the liquid state.

Throughout the first few million years the earth passed through a period of rapid change. Heat was generated by radioactive elements within its mass and by the force of gravity, that caused all matter to fall toward the center and form a more compact sphere. The solid materials began to melt. Dense ones like iron flowed inward and formed the metallic core. The less compact substances, containing a large fraction of silicon, oxygen, and aluminum compounds, floated to the top. There cooling occurred and the minerals solidified into *igneous* rocks. When the cooling took place below the surface it happened slowly. In these protected inside places large crystals formed, making the rocks known as granite and diorite and gabbro. When magma broke through the surface as lava, it solidified quickly in colder temperatures, forming finer-grained rocks: rhyolite, andesite, and basalt. These are composed of tiny microscopic crystals because there was not sufficient time available for the construction of large crystals. When the cooling took place very rapidly, crystallization could not occur at all. Then substances like volcanic glass and obsidian were produced. In fact, glass is silica that has been melted and quenched so quickly that it has not been allowed to crystallize.

Having made a firm, durable skin for the planet, nature might have rested from its labors. But this was just the beginning of a seemingly endless process of formation. Those very solid-looking rocks have been undergoing changes since the day they were born. Beaten by rain and wind, many igneous rocks have been broken down, altered by chemical reactions, and reassembled into *sedimentary* rocks. Volcanic action has heated some rocks deep below the surface, bent them into strange contorted shapes, melted them so the original crystal structure was lost. New crystals formed as the minerals cooled, making *metamorphic* rocks. Marble, for example, is limestone that has been heated and recrystallized.

These changes occur because something has happened to the rock; it has been cooked or squeezed or broken by outside forces acting upon it. But other much more surprising changes are taking place all the time. The chemistry is gradually altered deep within the rock and the internal structure is constantly being transformed. Wherever possible nature gets on with the job of building its favorite art form — the crystal. The amorphous solids like volcanic glass, obsidian, and several semiprecious stones, beautiful and useful as they may be, are relatively transitory manifestations of matter.

❧

Opals are among the most charming of semiprecious stones, with a luster like pearls and iridescent colors, soft and translucent. But the opal is not built of crystals. Composed of silica and water, it is one of those minerals that form in cool temperatures.

An interesting thing happens with the opal that tells something important about the way nature works with space and time, Very, very gradually, over millions of years, the opal loses its water content and crystallizes. Deprived of sufficient time "at the beginning" to weave the intricate design of the crystal, nature does not give up. Working throughout millions of years inside the extremely confined spaces between the molecules of the solid opal, it builds the quartz crystal at last.

Obsidian has a similar origin and suffers a similar fate. A smooth, glassy material, it is black as midnight and remarkably free from bubbles and other imperfections. This rock was highly prized by primitive people. In prehistoric times it was traded throughout Europe. Obsidian from deposits found in Central America were carried as far south as Peru and as far north as the Ohio River valley. It was chipped to produce razor-sharp tools. The Aztec priests used it for sacrificial knives to cut out the hearts of their living victims.

Obsidian usually forms on the outer skin of a lava flow where the most rapid cooling takes place. As it ages, the obsidian undergoes transformation. Beginning in tiny centers scattered rather uniformly throughout the rock, small white crystals form, creating "snowflake obsidian." Unaltered obsidian becomes progressively rarer in the rock formations as one goes back in geologic time. It is never found in deposits older than a few million years.

Other types of volcanic glass alter successively through various crystalline rock forms called zeolites and finally become feldspar, a mineral containing silicon, oxygen, and aluminum, combined with sodium, potassium, or calcium. Feldspar is more abundant than quartz in the earth's crust. No volcanic glass older than thirty million years has been found, and the intermediate zeolites are never older than 350 million years. On the other hand, many very ancient rocks are enriched with feldspar, which appears to be the end product of a transformation process so deliberate that 300 or 400 million years are required for its completion. Thus alteration continues even after a crystal structure has been formed, building more and more stable and balanced configurations of matter.

Many extensive layers of dolomite occur throughout the earth's crust. Yet geologists have looked in vain for places where the formation of dolomite is actually occurring in any significant quantity today. It appears to be the result of chemical changes that take place in a very leisurely way in the parent rock as water percolates through it. Limestone is gradually converted to dolomite — a harder and more stable stone — with the substitution of atoms of magnesium for every other atom of calcium in the crystalline structure.

The remarkable nature of the gradual change of one crystalline form to another without disturbing the shape of the whole is difficult to imagine. There is no human activity that is quite comparable. To visualize what this means, imagine that you are a child building a house of cards with many rooms and stories one upon another. Suppose that you do not have a complete pack of cards to start with, just half a pack with more hearts than spades, more diamonds than clubs. For your house you have in mind a nice precise pattern. You want to alternate the colors and the suits — a diamond, a spade, a heart, a club — and this scheme works out well for a while until you begin to run out of clubs and then out of spades. Still, you make do with what you have and put the two top stories on with all red cards. But the house doesn't really please you. A little later you come upon the other part of the deck and you go back to try to straighten out the design. Now the obvious way to do this is to tear the first house down and start over again. But the remarkable way nature undertakes this project is to leave the existing structure intact and replace each incorrect card individually without disturbing the whole. Remove the jack of hearts and put in the jack of spades and so on until the perfect pattern is complete. What legerdemain! It seems like magic. Yet many of the stones and rocks in the crust of the earth have been altered in some degree by this process.

A large rock formation in eastern Africa that has been dated as 500 million years old is composed of dolomite with layers of feldspar crystals. Geologists speculate that it is the end product of replacement processes that began at the margin of an ancient sea in which volcanoes were active. The continents in those early days were still devoid of life, but many organisms had begun to inhabit the seas. Their remains were washed up on the shores and were deposited in the shallow waters. From the compacted masses of these tiny shells, limestone formed. Wind-transported tephra from volcanic eruptions also accumulated on the water surface and settled to the bottom, making layers within the limy shell deposits. During the next half a billion years while vegetation took hold and bloomed upon the land, while reptiles and mammals and birds evolved, while the continents drifted apart and new oceans were created, the alchemist patiently at work within the rocks turned the crystals of limestone to dolomite and the little tephra particles to the translucent pink and white crystals of feldspar.

This formation is but one out of thousands of examples where we can trace the transmutation of the minerals in the earth's crust throughout geologic time. Until very recently the origin of the red and yellow colors seen in many rocks and soils around the world had not been identified. Several theories were proposed but none seemed to fit all the facts. Now we know that this coloring results from an alteration process. Like a slowly developing photographic print, the hue gradually appears and deepens as iron-rich minerals in basalt are weathered out, and — if the environment contains abundant oxygen and nonacidic groundwaters — iron oxides crystallize, coloring the rocks with many variegated shades of yellow, ochre, and deep rust red. Evidence of this process has been observed on the coast of Baja California, and dating has shown that the full development of the red color there took about five million years.

Transformations of this kind occur on a time scale so vast that we are scarcely aware of them. (Does a mite on a redwood tree sense that the tree is changing beneath his feet?) It is only when we see certain striking examples of the results of alteration that we pause for a moment and are amazed.

∾

There is a place in northeastern Arizona where the yellow and orange and rust-colored dunes of the Painted Desert dip down into a shallow basin that is strewn with strange, assorted piles of stone. At first sight it looks like a ruined city with huge columns lying broken and scattered on the ground. On closer inspection it becomes obvious that this is the graveyard of an ancient forest. Trees died here but never decayed; their remains turned into semiprecious stones.

An unusual set of circumstances provided the long period of geologic time that nature needs to perform this transformation. Organic substances usually decay much too quickly to allow such changes to take place. About 180 million years ago this land was part of a vast floodplain crossed by many streams. Tall conifers similar to our pine trees grew on the banks of the rivers. When the trees fell, they were buried by sand and mud from the overflowing streams that sealed off all air and oxygen from the deposits containing the fallen logs. Many years later volcanoes erupted nearby and a heavy layer of ash blanketed the region, just as the city of Pompeii was buried. But in this forest graveyard nothing happened to disturb the tomb for more than 100 million years. As in the case of the geode, water seeped through the overlying layer of rock, which was rich in silica. Molecule by molecule, quartz crystals formed in the tiny spaces within the wood tissue. In many specimens the cellular form was not disturbed when this crystallization occurred, and the precise pattern of the living tree was preserved down to the finest detail. We can even count the growth rings of trees that lived 180 million years ago.

In the volcanic environment of this graveyard, many elements besides silicon and oxygen were present. So other ingredients were added to the silica crystals: a bit of iron here, a pinch of magnesium there, a little titanium or chromium. From this palette emerged blood-red jasper, translucent green chrysoprase, carnelian and agate in vivid banded patterns of white, pale blue, or rusty brown. The colors pick out and accentuate the rings and cells of the living tree. In some places cracks and hollows in the original logs left open spaces. Here clusters of beautiful large crystals, clear quartz and amethyst, grew.

About thirty million years ago this part of the continental crust was lifted up in a great tectonic upheaval. As the land buckled beneath them, the petrified logs broke up. The sand and shale of the strata in which the trees lay had been painted with red and ochre of iron oxides as the volcanic tephra weathered, dissolved, and recrystallized. Then layers of the sediment were washed away except where it was protected by more resistant soil, making the striking flat-topped hills and buttes of the Painted Desert. Exposed again, their soft bed stripped away, the petrified logs rolled and tumbled until they came to rest in the bottom of the valleys and little gullies.

Many treasure troves like the Petrified Forest lie hidden beneath the earth's surface. Occasionally they come to light in a mine or a gravel pit or a river canyon. A petrified ammonite shell about 85 million years old was found in South Dakota. The chambered spiral of the shell is outlined in bands of blue, rose, and chartreuse crystals. Looking at this finely wrought piece of natural artistry, a strange thought occurs to me. Could human bodies be petrified like the ammonite? If Pompeii had not been excavated and had lain buried for millions of years, would the corpses of men and women and their pet dogs have been gradually turned into agate and jasper, or chrysoprase and amethyst? This would be the ultimate preservation of the dead in a form more suitable to eternity than the shriveled mummies of ancient Egypt and Peru.

༄

The process of alteration has provided earth scientists with a vast storehouse of information about the history of the planet. The shells and the shapes of soft-bodied creatures, even the tracks made by small scurrying worms that dug burrows in the primeval mud to escape the fierce armored jaws of the huge fish that dominated the very ancient seas — these impressions have been gradually hardened, made denser, more resistant to the vicissitudes of time. Most shells when first cast off by the organism that made them are light and fragile. They can be easily crushed by the pressure of a finger. But the ancient ones we find buried in the earth, enclosed in sandstone or limestone or shale, are usually hard and heavy, strong enough to carry down through many millennia information about the world in which these organisms lived and died.

Masses of ancient algae called stromatolites have been found in formations more than three billion years old. These algae lived in warm shallow waters long before the first land plants evolved. The continents were bare as the windswept surface of Mars and breezes stirred the surface of the seas, making ripple patterns that are still preserved along with the remains of these very early plants, arranged in layers, long and wavy like mermaid's hair.

As I run my finger over the fossil markings in these primeval rocks, I am reminded of Mark Twain's comment: "There is something fascinating about science," he said. "One gets such wholesale returns of conjecture out of such trifling investment of fact." And yet as one sees rock after rock and becomes familiar with the distinctive impressions made so long ago, it becomes obvious that information is inscribed there just as surely as it is conveyed by the marks on this page.

With infinite pains paleontologists are deciphering the code in which the history of the earth is written. Many dates and facts are coming to light as the symbols are read, but the meaning of the whole is still obscure. Something of incalculable importance is being accomplished with timeless patience as, crystal by crystal, quartz is replacing organic matter in the deep interior of the dead treetrunk. Molecule is carefully balanced upon molecule to make the violet-colored amethyst in the enclosed heart of a geode. On the cosmic scale nature is busy even today distributing new stars in the firmament, sweeping more space and time into an ever-expanding pattern. Nothing in the universe is really finished. Everything is in a state of metamorphosis and has been for at least eighteen billion years.

But the nature of this cosmic enterprise is still as much a mystery as it was before the dawn of human thought. It may even have troubled Lucy, that very distant ancestor of ours who lived on the shores of a long-vanished lake in northern Africa three million years ago. One day she might have picked up a handy stone on the mountainside near her home. The rock was rough and fitted neatly into the hollow of her hand so she could use it as a hammer to crush the nuts she had gathered from the trees on the riverbank. At the first blow the geode split open and beauty spilled out. Looking into the brilliant pattern of white light and violet darkness, Lucy may have been reminded of the sky at dusk with stars just beginning to come out. Wonder flickered briefly behind her simian brow. Why do the spirits of the evening sky and stars hide inside a small dark rock? The answer eludes us even now. The secret still lies sealed within the stone.

10

Messages Written in the Sand

Who can number the sand of the sea,
and the drops of rain, and the days of eternity?

— ECCLESIASTICUS 1: 2

There is no better way to understand the slow pulse of geologic time than to spend a day on one of the lovely beaches of the earth — to dig your toes in the warm sand, bury your hands deep in the golden heaps of pure earth-stuff, and let the grains flow in a shining stream between your fingers. Or lie for an hour cradled in its soft lap, suspended between earth and sun, soothed by the sight of rhythmically rolling dunes against the blue sky like wind waves in a field of ripe wheat if time could be made to stand still. And indeed the bird of time does seem to fold its wings and come to rest upon the sand. The tiny fragments running through your fingers are more ancient than the soaring mountain ranges or the rocks that lift dark shining whale-backs out of the sea. A grain of quartz sand is as near to being eternal as any thing on the earth's surface.

Along the hard beach at the water's edge, wet from the ceaseless to and fro of the sea, the retreating waves leave crescents of foam — tiny bubbles not much larger than a grain of sand but each one holding miniature rainbows, each one lasting but a moment before it breaks in a sparkle of light and sound. If you listen very carefully, you can hear the bubbles fizz as in a glass of champagne. Here on this little stretch of earth the most ephemeral and the most eternal lie together in closest harmony as if to mock the tyranny of time.

The sand that covers most of the beaches as well as the deserts of the world is composed of little translucent quartz crystals. Released by erosion from the great continental rocks, these countless trillions of tiny pieces have been winnowed, sorted, refined, and polished by the wind and the waves. As long as the breezes have blown, rain has fallen upon the land, and ocean currents have flowed around the planet, sand has been born grain by grain from the rocks of the crust. And once a sand grain has been created, only an exceptional combination of circumstances will destroy it. The quartz crystal is so hard and resistant to attack from the usual natural forces that it is one of the longest-lasting materials on earth.

~

The granitic rocks of the continental crust are composed principally of quartz and feldspar crystals. Granite (a name derived from the grainy texture of these rocks) can be crumbled to bits by wind and rain and flowing water as these forces act on it throughout millions of years. One of the reasons for its susceptibility to weathering results from a peculiarity in the shape of quartz crystals. When they first form from hot magma, the atoms are arranged in a way that takes up a maximum amount of space. Then as the rock cools further, the crystals change their shape, becoming slightly smaller and denser. But rearrangement is constrained by the surrounding rock. Stresses develop between the crystals, and they can be broken apart more easily. The small fragments released by physical and chemical weathering become sand grains.

The large pink and white grains that usually occur in granite are feldspar crystals. Less stable than quartz, they dissolve and react with relative ease. And when subjected to the action of running water they become mud and clay that may later be compacted by overlying materials and harden into shale.

Granitic rocks contain small amounts of other minerals in addition to quartz and feldspar — red garnet, black magnetite, translucent mica, dark biotite, white zircon, and many-colored tourmaline, to name just a few. These crystals, too, are weathered out of the rock, broken down into tiny pieces, and transported by the same processes that move quartz sand. They turn up on many of the beaches around the world, forming little streaks and patches, windrows making faint ripples on the pale sand dunes. Where they are more thinly distributed, they seem to deepen the shadows on the beach or add a glitter like diamond dust.

In some places these less abundant minerals may comprise the predominant sand, as on the beaches at Bristol Bay in Alaska, which are made up of almost pure magnetite, or on the many somber black Hawaiian beaches that are composed of olivine derived from black basaltic rocks. The lead-gray beaches of Saint Simons Island in Georgia are heavily laced with ilmenite and rutile.

In tropical waters where marine life is abundant and reefs lie close offshore, beach sands are often made up of pulverized coral fragments mingled with the remains of calcareous seashells. Pick up a handful of these sands and you may find miniature shells still intact.

Coral sand beaches, tinted various delicate shades of pink, accentuating the blue of the ocean and sky, are among the most beautiful beaches in the world. But compared with the ancient quartz sands, they are recent in origin. After the continental rocks formed, several billion years passed before the earliest organisms with calcium carbonate shells appeared in the oceans. Untold numbers of these had to live and die to produce a thick accumulation of shells on the seafloor, and more time had to pass before they were crushed by the pounding of the waves and the tides, which gradually separated out the lightest fragments and washed them ashore.

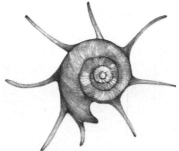

Coral sands are not as thoroughly smoothed and rounded as most quartz sands are. They tend to cling like tiny burrs to beach towels and sunbathers' bodies. They are also cooler to the touch, as though they brought with them the chill of the ocean depths from which they sprang.

Quartz crystals are much more efficient than coral in absorbing and holding the heat of the sun. At midday the silica sand beach may be too hot to walk across with bare feet. I remember a wide crescent-shaped beach at Kovalam on the southwestern coast of India where golden sand was piled into soft inviting-looking dunes, tempting as an eiderdown. But I soon found that they were too hot to provide a comfortable resting place. After eleven o'clock in the morning on an early spring day it was necessary to pass quickly across these dunes to the safety of the water-cooled sand at the ocean's edge. Even in Arctic regions sand dunes often heat up to temperatures surpassing 100° F in summer and people exploring these Arctic deserts find it wise to stay close to streams in order to cool sand-toasted feet.

Time is the essential factor in the making of sand — time for the slow but persistent forces of erosion and weathering to remove the grains from the rocks, time to move them from the forest to the sea, from the mountaintop to the desert dune. A river needs about a million years to transport an average-sized grain of sand a hundred miles downstream. The lightest grains are carried in suspension; they move the farthest in any given time. The heavy ones drop to the streambed where they roll and bounce slowly along the bottom. Thus a crude sorting by size takes place and the farther the sand is transported from its original source, the more thoroughly the grains are separated by size and weight. When they reach the sea they are subjected to the vigorous action of the waves and the tides. Eventually they may be left by a retreating tide on the water's edge, picked up by the wind, and deposited on a beach.

But even then the odyssey of the sand grain has just begun. It has passed through only one cycle of sorting and winnowing. Since the beginning of the world these same materials have been worked and reworked many times. Ancient beaches and desert dunes have been compacted and cemented into sandstones. With changing land and sea levels, erosion has removed the overlying material, broken up the sandstones, separated out the individual grains again, spread them in new patterns across the earth, incorporated them into other stones, eroded the stones, and sorted out the grains once more. The most venerable sands have passed through hundreds of these cycles. Each time the sand is reworked it is further refined, polished, and grouped by size and kind. Nature seems never quite satisfied with the purity of the product and must try once again.

Since quartz crystals have a characteristic spatial configuration, grains of sand are remarkably similar around the world whether they come from the Sahara Desert or Palm Beach or the Great Kobuk dunes of the Arctic. But some differences in the size and shape of quartz crystals do exist and, furthermore, all sand in nature contains admixtures of other minerals. These variations tell an interesting story about the life history of that particular deposit.

If a sample of sand is taken to the laboratory, it can be sifted through finely graded sieves and the grains falling in each size range weighed. The size distribution reveals the kind of sorting process through which the sample has passed. Sands carried and deposited by glaciers are the most poorly sorted, containing the greatest range of sizes and many mineral particles other than quartz. Glaciers are not selective; they pick up and transport anything in their path. Sands deposited by rivers usually contain many fine particles of mud and silt as well as quartz crystals. Sands deposited on the edge of the sea where breaking waves continuously wash thin sheets of water back and forth never contain fine-grained sediments that are light enough to be suspended in the water. The rapid movement of the waves does not allow them to come to rest on the bottom. This oscillating motion produces a very thorough sorting action. Wind is also an effective harvester of finely graded crystals. Constantly winnowing the sand, it picks up small-grained sediments and blows them away. So desert sand is frequently very pure quartz particles without admixtures of other minerals.

The selective action of air movement is dramatically displayed in desert sand and dust storms. Even on days when there is little or no wind, the air in contact with the sand is heated and rises, carrying the very fine dust particles into the atmosphere. On extremely hot days these "dry fogs" may reduce visibility to a dangerous level, and the sun is often obscured for many hours. When a light wind blows, it picks up heavier dust and bears it aloft. Dust particles differ greatly in size; so a dust cloud does not have a clearly defined upper boundary. The smallest particles may soar thousands of feet high and not come to rest again for many months. Dust that originated in the arid areas of North Africa is carried on the trade winds out into the Atlantic Ocean. It has been found even as far west as Barbados.

The dust storms that characteristically occur on the leading edge of a cold front are among the most terrifying spectacles in nature. The cloud can be seen approaching across the flat wastes of the desert. A tremendous, impenetrable yellow-brown mass suddenly blots out the sun. The desert traveler who is experienced in these storms covers his nose and mouth with wet cloths to avoid being suffocated. As the center of the storm passes, darkness as dense as night can prevail.

These clouds consist mostly of light dust particles, not sand. Wind must be stronger to move the sand grains. When it increases in speed, narrow ribbons of sand begin to flow just an inch or so above the ground. Soon the ribbons widen and merge, forming an unbroken sheet of sand moving swiftly over the surface.

A German naturalist, Uwe George, has given us an eyewitness account of such a storm in the interior of the Sahara, near the oasis of In Salah in Algeria:

> First came the moderate winds and tremendous drifting masses of dust, which gradually changed to a dense carpet of sand hovering over the ground from horizon to horizon. The wind became stronger. It blew strongly and evenly, and soon the air seemed to consist of nothing but yellow grains of sand, which rattled against the windshield of our car as if someone were throwing shovelfuls of sand. At times visibility dropped to zero. Before everything was blotted out, I had noticed that the track I was following ascended a rise several hundred yards away. I drove straight on until I reached that hill, and the effect was like that of a submarine suddenly breaking through the surface of the water, for abruptly the car rose above the upper limit of the sandstorm. The spectacle I beheld was unique. I stood under a blazing sky and looked down at the top of a seemingly infinite sea of sand flowing along at high speed. The brow of the hill on which I stood formed an island in the midst of this surging golden yellow sea. At some distance I could make out the brows of other hills, other "islands." But the most curious part of the experience was the sight of the heads and humps of several camels that appeared to be floating on the surface of the sand, like ducks drifting in the current of a river. The camels' heads and humps, all that showed above the upper limit of the sandstorm, were moving in the same direction as the drifting sand, but a great deal more slowly.

An abrupt, sharp upper limit is typical of sandstorms, for the reason that the storms are so often formed by very steady winds. Unlike particles of dust, grains of sand are almost all of about the same size and weight. A specific wind velocity will raise the uniform grains of sand to a uniform height. . . . Very strong winds can raise the sand to a height of about six and a half feet. . . . No reliable statistics are available, but there is indirect proof of the assumption that the sand whirled up by sandstorms usually reaches that height. It is the height of camels, which have become adapted to the desert. The heads of these beasts reach just above the densest zone of drifting sand, giving them a prime requisite for survival in the desert: the ability to breathe air that is relatively free of sand.

Under unusual circumstances when wind velocities approach hurricane speeds, sand is carried to greater heights and begins to circle in the form of tremendous whirlwinds. One after another these swirling clouds follow each other, sometimes continuing for days. Sand driven at such speeds is dangerously abrasive, like a sand-blasting machine. Air temperatures rise and humidity drops to a few percent. Anyone caught out in such a storm is in grave peril. Many bodies of sandstorm victims have been found in the desert, completely dried out like mummies.

Even these most severe earthly sandstorms are small and of short duration compared with the storms that have been observed on our neighboring planet, Mars. Once a year, usually at the end of the Martian spring when the planet is at its closest approach to the sun, a shining white trail extending for thousands of miles appears in the Southern Hemisphere. Within several days it darkens and turns yellow or red in color. Spreading rapidly, it moves in a westerly direction, circles the globe, and envelops the whole Southern Hemisphere. Sometimes it flows over into the Northern Hemisphere and within a few weeks it has involved the entire planet. When Mariner 9 arrived in the vicinity of Mars, it observed such a storm, which totally obscured the planet for weeks. It is estimated that wind velocities in these storms must exceed a hundred miles an hour. Sand carried on winds of this strength without trees or other vegetation to brake its violence would make a sandstorm on the Sahara seem as insignificant as a little dust devil rising from a plowed field on a warm spring day.

Many of the photographs of Mars show drifts of windblown sand and the largest sand desert found anywhere in the solar system surrounds the frozen north polar cap of Mars. From this evidence we know that the winnowing, grading, and separating of planetary material into heaps of nearly identical crystals is not just a peculiarity of the earth. It may be a cosmic phenomenon.

The dune fields surrounding the Martian poles were probably formed by strong winds transporting and redistributing particles left by retreating ice caps. A similar process has made sand deserts in the polar regions of the earth. The Great Kobuk dunes in Alaska are the remains of an ancient desert that covered at least 300 square miles during one of the last ice ages, 33,000 years ago. Outwash from the glacier was picked up, sorted, rearranged, and deposited according to size and weight by the winds.

The interplay of sand and air throughout long periods of time is written in the shape of the dunes on the great sand deserts of the earth. Like the sea, their surfaces are ruffled by the movement of air masses, but the heavy medium of sand responds much more slowly than the mobile water surface. The patterns flow and change almost imperceptibly; so at any one moment the desert looks like a frozen sea stretching from horizon to horizon. Dunes do migrate downwind, however, at rates of several inches a day. Time-lapse photography would show the sand surface to be in constant turbulent motion, a pale-colored sea piled into long rollers by the steady sweep of the prevailing winds, tossed into steep breakers by storms, stirred into ripples by a thousand vagrant breezes.

Longitudinal dunes, or seifs, are created by the prevailing winds blowing across deep deposits of sand; their cylindrical shapes extend in parallel waves sometimes as long as one hundred miles and taller than twenty-story buildings. Smaller, crescent-shaped waves of sand, called barchans, march in serried ranks, the open sides of their sickles facing downwind. This distinctive shape is caused by winds blowing persistently from the same direction in places where accumulations of sand are not very deep. The center of each barchan is larger and heavier than the edges; it moves more slowly than the sand at the edges so the sides are blown out ahead and form the horns of the crescent. The concave slip faces are very steep. Sand grains carried over the smooth upwind side of the dune fall onto the face, steepening its slope to the point where grains can no longer rest there, and cascade downhill.

Strong variable winds pile up high dunes and troughs like breakers in a stormy sea. In low, partially protected places where the air currents are diverted so they flow inward from all quarters, star-shaped dunes often form and rise to amazing heights. They are composed of ridges rising to a central peak sometimes a thousand feet above the desert floor. These dunes do not migrate like the others but remain fixed in one location for many decades.

Across the long surfaces of the seifs, along the gentle windward slopes of the barchans, and on the stepped terraces of the star dunes are impressed row after row of tiny ripples like those that dimple the ocean surface when a light breeze begins to blow.

The forbidding immensity of the great sand seas can be understood only by those who have crossed them on foot and camelback. A British traveler, Wilfred Thesiger, described his crossing of the dunes called the Uruq al Shaiba in the Empty Quarter in Arabia:

> A high unbroken dune-chain stretched across our view. It was not of uniform height, but like a mountain range, consisted of peaks and connecting passes. Several of the summits appeared to be seven hundred feet above the salt-flat on which we stood. The southern face confronting us was very steep, which meant that this was the lee side to the prevailing winds. . . . It seemed fantastic that this great rampart which shut out half the sky could be made of wind-blown sand. . . . A succession of great faces flowed down in unruffled sheets of sand, from the top to the very bottom of the dune. They were unscalable, for the sand was poised always on the verge of avalanching, but they were flanked by ridges where the sand was firmer and the inclines easier. It was possible to force a circuitous way up these slopes, but not all were practicable for camels, and from below it was difficult to judge their steepness. . . . We led the trembling, hesitating animals upward along great sweeping ridges where the knife-edged crests crumbled beneath our feet. Although it was killing work, my companions were always gentle and infinitely patient. The sun was scorching hot and I felt empty, sick, and dizzy. As I struggled up the slope, knee-deep in shifting sand, my heart thumped wildly and my thirst grew worse. I found it difficult to swallow; even my ears felt blocked, and yet I knew that it would be many intolerable hours before I could drink. . . . It took us three hours to cross this range.

Travelers in the desert have often reported hearing strange sounds varying from a soft, persistent hissing to sudden high-pitched musical sounds or roars. The British geologist R. A. Bagnold, studying dunes in southwestern Egypt, reported that on two occasions during still nights a vibrant booming suddenly broke the stillness, "so loud that I had to shout to be heard by my companion. Soon other sources, set going by the disturbance, joined their music to the first, with so close a note that a slow beat was clearly recognized. This weird chorus went on for more than five minutes continuously before silence returned and the ground ceased to tremble." The Egyptian geologist Farouk El-Baz described similar reverberating sounds as well as softer "singing" noises in Egypt's Great Sand Sea. He suggested that they are caused by the movement of sand particles as the dune shifts in the direction of the wind. Steady motion generates a soft sibilant sound as the sand grains climb the windward face, pause for a moment at the dune crests, and cascade down the steep slope. More sudden shifts cause avalanches, generating the louder sounds. Although the hypothesis has not been proved, this is one possible explanation of the "singing sands of the Sahara."

Men who have crossed the deserts have been surprised and, at first, mystified by another strange phenomenon. Trails of camel footprints have been observed leading up one side of a dune, reaching the summit, and then suddenly disappearing as though the camel had taken wings. In the distance the trail materializes again, threading its way up the next dune, and vanishing at the crest. A moment's thought dispels the mystery. The downwind sides of the dunes are so steep that avalanching occurs as the animal descends, covering his trail as neatly as if a fresh snow had fallen.

∽

Like the wind, moving water impresses its characteristic patterns on sand surfaces. The one-way flow of a river creates parallel rows of asymmetric ridges on the riverbed. As in the windblown dunes, the steep sides of the ripples face downcurrent and the whole pattern migrates slowly in the direction of flow. On the edge of the sea the back-and-forth water movements of surf and tide mold sand surfaces into more symmetrical corrugations.

The speed of flowing water also influences the shape of sand ripples in ways that can be easily identified. Delicate, parallel laminations indicate gentle water movement. As the current speed increases, the ripples become larger, their pattern more confused and interwoven. In shallow waters the buildup of sand waves affects the movement of the surface water. Breakers and troughs develop that in turn cause turbulence on the bottom and alter the sand waves. All of these signs can be read, yielding much information about the forces that have molded that particular pattern in the sand.

Throughout the lifetime of the earth the shapes of many dunes and beaches have been preserved in layers of sandstone. The waves and ripples, still imprinted there, tell us which way the wind was blowing, where the rivers flowed, and how the tides moved along the shores hundreds of millions, even billions, of years ago. The quartz crystals themselves do not reveal their age but index fossils and remains of organic materials embedded in the rock provide the necessary clues. Sand was already being washed from continental rock in very large quantities about two and a half billion years ago. Some of the ancient rocks show ripple marks characteristic of beaches or shallow offshore waters. In other deposits thick beds of sandstone typical of desert formations were apparently laid down during very dry periods in the earth's history when enormous amounts of sand were moved by the wind. The great fossil dune beds of Permian (250 million B.P.) and Jurassic (150 million B.P.) age bear witness to eras when the transport and deposition of sand dwarfed anything found on earth today.

During the last half-century, however, the earth's deserts have been increasing very rapidly in size — not only the sand-covered lands but also the vast areas of rocky and sun-scorched soil that are too dry to support human life. Since 1930, some 450,000 square miles have been added to the Sahara. Tunisia has lost almost half of its arable land. In the Sudan the desert boundary has moved southward sixty miles in just the last two decades and the northern Sahara is also increasing at an inexorable pace, adding an estimated 250,000 acres to the desert every year.

Natural forces like those that caused the fossil dune beds of the Permian and Jurassic may be responsible for some of this steady increase in arid lands. But mankind is also contributing to the process of "desertification." The growing pressure of human population causes farming, grazing, and fuel gathering to be extended into marginal regions. As the protective covering of vegetation is reduced, the land becomes more susceptible to erosion and desiccation by the winds. The area suitable for grazing shrinks, and at the same time the number of animals that must be raised increases because human populations are growing rapidly. The destructive pressures on the delicate ecosystem of arid lands are reaching unprecedented proportions as almost five billion human beings try to scratch a living from the earth.

By agricultural reforms and good land management, marginal lands can be spared desertification. By heroic measures, such as the installation of irrigation systems, some desert lands can be reclaimed. But those lands covered by sand or downwind of sand deserts are the most difficult to spare or to reclaim. Between Cairo and the Aswan Dam sand dunes moved by the prevailing winds are encroaching on the fertile Nile valley. That precious irrigated portion is only 4 percent of Egypt's total land; the rest is desert. In Saudi Arabia attempts have been made to arrest the movement of the dunes by spraying the sand with a grain-cementing material — a futile human gesture at controlling the forces of nature. As one flies over the seemingly endless expanses of the Sahara Desert one can appreciate the hopelessness of attempting to cement together these innumerable grains of earth crust — to arrest even for a decade or two the inexorable movement of the shifting dunes. The sands never rest; gently but surely they invade, destroy the life forms that have maintained a precarious hold on that portion of the planet's surface, and reestablish the supremacy of earth crust.

 ❧

Sand in quantity is awesome. The sheer multiplicity of it humbles the imagination. "Who can number the sand of the sea . . . and the days of eternity?" The single grain, on the other hand, seems to be the perfect model of insignificance, a crumb of matter possessing no individual worth, no value except as one more unit in the vast collective immensity. But pick up one of these tiny quartz crystals. Turn it over in your hand, study its surface and the ways it reflects the light. It will soon be apparent that the single sand grain does have distinguishing features and an individuality acquired in its long journey through geologic time. All the vicissitudes through which it has passed have left their marks upon its surface like wrinkles and scars on an old man's face.

Grains of desert sands, which are constantly winnowed by the winds, are usually well rounded with a frosted surface like ground glass. The frosting is caused, at least in part, by abrasion with other sand grains. Quartz crystals on beaches and in riverbeds are not subjected to the same degree of sand-blasting as the desert grains. The individual crystals often have polished, shiny surfaces.

When grains of sand from various environments are examined under an electron microscope, characteristic surface features can sometimes be identified, providing clues to their life histories. Deposits as old as 300 million years still carry detailed information about their past. Sand grains that have been moved by glaciers bear the marks of surface breakage caused by the heavy grinding of the ice. Sand crystals from all aqueous environments — rivers, beaches, continental shelves, and deep sea basins — are frequently inscribed with little pits and V-shaped depressions. These occur only sparsely on river sands; about 15 percent of the grains are pitted in any way. Beach sands from rather quiet coastal environments carry marks on a slightly larger percentage of the grains, but sand from more turbulent coasts bears the V-shaped impressions on more than half of the crystals. And sand that has been carried over the edge of the continental shelves by swift underwater currents is marked on almost a hundred percent of the grains with these distinctive traces.

By reading the surface marks and noting how they lie in relation to one another we can reconstruct the life story of that grain of sand — its origin and the adventures through which it has passed. We can tell, for example, that a grain of sand had been carried by a glacier from the place where it was eroded from continental rock, that it was later released from the ice, and transported by a stream to the sea where it was tossed and rolled about in the surf, thrown up on a beach, picked up by the wind, blown across desert dunes over a long period of time, until it was deposited again in coastal waters, carried by underwater currents over the continental shelf, and finally came to rest in a pile of mud and silt on the ocean bottom. So the odyssey of the single grain of sand is inscribed on its face like the Lord's Prayer written on the head of a pin.

❧

The noted French mathematician Pierre Laplace declared that the ultimate aim of science is to demonstrate from "a single grain of sand the mechanics of the whole universe." "All events," he said, "even those which on account of their insignificance do not seem to follow the great laws of nature, are a result of it just as necessarily as the revolutions of the sun."

The insignificant grains of sand on the beaches of the earth and the deserts of Mars do carry an interesting message about the workings of the universe. There is a principle in physics known by the forbidding title of the Second Law of Thermodynamics. It states that molecular disorder increases in all spontaneous changes. For example, if a drop of red dye is added to a glass of water, the molecules of the dye spread out, distributing themselves more and more evenly among the molecules of water until the two are thoroughly mixed. Where originally we had water and dye we now have dyed water. "Order" in the sense of a high degree of differentiation has decreased and considerable effort would be required to restore that order.

The British physicist James Clerk Maxwell made up an imaginary scenario to illustrate this principle. In order to separate the molecules out again you could take two containers connected by a narrow tube and in the tube you would have to place a Little Demon to direct the flow of traffic, allowing only red molecules to pass from right to left through the tube and only colorless molecules to pass from left to right. In this way the dye and the water could be separated out again and molecular order would have increased.

The Second Law says that there are no Little Demons directing traffic in spontaneous natural processes. The molecules might be separated again by interfering with nature — by distillation or by chemical reactions — but the orderly arrangement achieved by these processes is brought about at the expense of increased disorder somewhere else. Left to itself, nature will tend to spread everything out into an undifferentiated and ultimately featureless world. Physicists have interpreted this to mean that the universe "is running downhill."

I had been quite thoroughly indoctrinated in this belief so I was very much surprised when I learned how sand has been distilled from the rocks of the planet. Beginning with the granitic crust where quartz, feldspar, garnet, mica, tourmaline, zircon, and many other kinds of crystals are mixed in a haphazard way throughout the fabric of the rock, natural processes have separated out the quartz crystals, sorted them by size, rounded and polished them, and collected them in vast lustrous piles. This distillation of the earth's crust has been achieved by the random action of wind and wave and rain since the birth of the planet. Proceeding inexorably, the sorting has become more perfect with the passage of time. Perhaps Maxwell's Little Demon exists, after all!

But the Second Law of Thermodynamics is one of the cornerstones of science. Its worth has been proved a million times over in problems dealing with the transfer of heat and energy. It cannot be disproved by a pile of sand. "You have shown me a local increase in order," the physicist would say. "Somewhere else disorder has increased — probably in the winds and ocean currents that move the sand. You must describe a totally isolated system in order to prove that there has been a net increase in order."

But how can we have a truly isolated system? As John Muir expressed it: "When we try to pick out anything by itself, we find it hitched to everything else in the universe."

The winds and the ocean currents move in response to the energy of the sun and the spinning of the earth; these are affected by gravity and the rotation of the galaxy, and these . . . In our search for something that can be considered complete in itself, we must look at larger and larger pieces of the cosmos until we come to the universe itself — the only truly isolated system.

And what about the universe? Is disorder increasing there? Is it relentlessly running downhill eon by eon, becoming steadily more undifferentiated and featureless? To answer this question let us take a look at its development throughout the eighteen billion years of its existence. According to the generally accepted theory the universe began with a cosmic egg — an enormously compacted lump containing all matter and energy. At time zero the egg exploded with a force so great that the pieces are still hurtling away from each other in all directions. In the first few minutes there was chaos beyond imagination as scraps of matter rocketed out into space. This intensely hot "plasma" expanded and the random forces of the explosion distributed it into clouds of varying density. Gradually the plasma cooled and simple atoms began to form. The matter dispersed in clouds began to come together under the influence of gravity, forming whirlpools and smaller eddies. Stars were born from the whirling clouds; galaxies took shape, organizing the stars into vast spinning wheels. Atoms came together, making more and more complex molecular assemblies. These large molecules floating in the cosmic clouds condensed and formed planets like our earth. And still the complexity of matter continued to increase. On the surface of the earth many types of organic molecules took shape and finally the molecule capable of reproducing itself, thus multiplying manyfold its own orderly arrangement.

This is not the life story of a universe that is running downhill. For the last eighteen billion years there has been a continuous evolutionary process building more highly structured, more finely differentiated kinds of matter. Beginning with a chaotic mass of fleeing plasma it has made the Milky Way and Andromeda, the Pleiades and the shining rings of Saturn. Like an artist tossing off one beautiful form after another but never quite satisfied, it has created the angel fish, the lotus flower, the ruby-throated hummingbird.

And year after year, eon after eon, it has tirelessly piled up heaps of golden sand. Refined and polished, these humble and insignificant crumbs of matter may be telling us something important about the nature of creation, something that transcends the Second Law and predicts a very different future for the cosmos. Just as the petrified ripples reveal the movement of the ancient river's current, so the great accumulations of purified earth crust may be pointing out the direction of universal change.

From the life history of a single grain to the flow of cosmic time — all these messages are written in the sand.

Buried Treasure

The earth is full of thy riches.
So is this great and wide sea . . .

— PSALM 104

The tropical island of Sri Lanka, green as an emerald cabochon set in the aquamarine waters of the Indian Ocean, is endowed with beauty and wealth far exceeding that of most lands. Coral beaches ring its shores; coconut groves and rice paddies clothe its lowlands; and cool mountains crown its center, providing the perfect conditions for mile after verdant mile of tea plantations. Pineapples, bananas, breadfruit, jackfruit, and cashews grow abundantly on the island. Fish and prawns abound in the ocean, the many lakes, and man-made reservoirs. Food is available just for the plucking, and precious gemstones lie scattered in the soil throughout the land. Rubies and star sapphires have been found lying in the mud of rice paddies. Cinderella fairy tales do come true in Sri Lanka.

The island lies near the equator and the air is soft and warm throughout the year. Abundant rainfall causes rapid weathering of the surface rocks and results in the production of a nutritious red clay soil that is ideal for the culture of tea. Swift-flowing rivers carry this surface soil down into the valleys, spread it out into fans of sediment that in many places is twenty or even forty feet thick. It is in these river valleys that the richest deposits of precious stones — rubies, sapphires, tourmalines, spinels, garnets, citrines, alexandrites, and zircons — have been discovered.

Ratnapura, the "city-of-gems," lies in such a valley nestled against densely vegetated hills on the edge of land where rice has been grown for untold generations. Beneath the deep layer of red clay lies a bed of gravel known as *illam*. It looks like quite ordinary gravel with many quartz and granite pebbles; only an expert can perceive that it contains an amazing array of gemstones.

The mining of these treasures is performed in a very primitive manner, virtually unchanged since the mines were first worked thousands of years ago. A hole is dug to the depth of the *illam* and the sides of the pit are reinforced with long vertical poles driven against the walls. Then branches and palm leaves are wedged behind them to hold back the soft clay. When water begins to collect in the bottom of the hole, it is bailed out to prevent flooding. The only modern innovation used in these digs is a water pump, which enables the miners to go to depths of forty feet. Even with the pump the miners are often working ankle-deep in muddy water. High humidity and tropical heat are intensified in the stagnant air at the bottom of the pit. The men who work there wear only loincloths and a heavy spattering of mud.

The mining operation proceeds at a leisurely pace with an undercurrent of excitement and anticipation, making it seem more like a sideshow than a serious undertaking. Children, relatives, and friends stop by to watch, hoping to witness a moment of great discovery.

The men in the pit fill their round bamboo baskets with gravel, dip them in the water, and swirl them expertly, causing the lighter sand, clay, and mud to float to the top and spill over the edge. The heavier pebbles, including any gemstones, collect on the bottom. Periodically the baskets are raised on pulleys to the surface where an experienced "sorter" examines the stones and sets aside any that look promising. The baskets are then emptied and lowered again into the pit. Watching this primitive procedure, one finds it difficult to imagine that many precious stones can be found in this way. Even the best of the pebbles picked out by the sorter look too dull to be potential jewels. Their surfaces are cloudy and diffused, obscuring the brilliant color which may lie beneath. But each year gems totalling hundreds of thousands of dollars are recovered from these mines. They were one of the world's most important sources of gems long before the Christian era.

Sapphires of many hues are the most valuable stones found in great abundance here. They may be yellow, pink, pomegranate, or tangerine, as well as the deep, clear blue most often associated with this stone. Rubies differ from sapphires only in the coloring agent. By convention the red or dark purple form of the mineral corundum is called ruby and all the other colored varieties are sapphires. Corundum, an oxide of aluminum, is a very dense, resistant mineral, second only to diamond in hardness. It is much more plentiful than diamond in the earth's crust, but fine, large crystals of gem quality are seldom found. Those that are discovered are among the most valuable stones in the world. Rubies derive their color from a trace of chromium replacing some of the aluminum atoms in the crystal structure. Traces of iron and titanium produce blue and yellow gems. Many glorious star sapphires with silky six-rayed streaks have been found in Sri Lanka and are the special pride of this country.

According to an ancient Asian myth the earth is carried on an enormous sapphire — the celestial stone — and the reflected light from this gem creates the blue sky. Today we can see the earth from space and the view is strangely reminiscent of this ancient myth. The planet looks like an uncut sapphire displayed against the velvet black of space — soft translucent blue with cloudy surfaces etched by the abrasion of billions of years across its rounded face.

~

Sapphires of very fine quality have been discovered at a number of sites in Asia — Kashmir, Burma, India, as well as Sri Lanka. The best rubies come from an area near Mandalay where they have been mined since the Stone Age. In most of these locations the gems have been retrieved from gravel beds by methods similar to those still practiced in Ratnapura.

In these placer deposits much of the work of mining precious stones has been accomplished already by the forces of nature. Raindrops pelting the hard rock surfaces year after year gradually dissolve the softer materials, leaving the most resistant kernels intact. When all these sediments are washed downstream, a sorting action takes place (just as in the case of sand). The heavier particles tend to fall toward the streambed and are separated from the lighter ones that float near the top. Eventually they are deposited in discrete layers along the river deltas.

Although placer deposits are relatively accessible and can be exploited with primitive techniques, all miners dream of finding the mother lode — an even more concentrated deposit of precious stones embedded in hard rock. This is a dream that has not been realized in Sri Lanka. With the exception of moonstone and garnet no gems have been found in the rocks where they were originally formed. Since most of this island is composed of very ancient rock that has been continuously eroded throughout hundreds of millions of years, we might conclude that the beautiful crystals emplaced in the rocks have all been released from their stone settings and carried downstream. But it is also possible that deeper veins still exist, concealed in the core of the granite mountains beneath the tea plantations.

Throughout the island it is easy to observe evidences of the massive volcanic action and tectonic movements that created the heat and pressure needed for the formation of gems. Swirling lines of light and dark layers mark the ancient cliffs, showing the results of forces great enough to bend solid rock as though it were putty. Great outcroppings of igneous rocks, the necks of several ancient volcanoes, make dramatic outlines against the horizon. Gleaming biotite and dark garnet chips stud the exposed rock surfaces.

In many places there are unusually large crystals of mica and quartz characteristic of pegmatite formations. These showy deposits are created by a process that causes a concentration of certain minerals. Magma forced up from the hot, soft layer of the earth beneath the lithosphere begins to cool and forms the familiar solid substances that make up volcanic rocks. The minerals that solidify first are those that crystallize at high temperatures, leaving the remaining liquid enriched in those minerals that crystallize at lower temperatures. As this process continues, different types of rock take shape: first the dark basaltic rocks, then the paler granitic rocks. The final liquid residue is a solution enriched in the light elements. When this solidifies in enclosed, protected places, it forms crystals of unusual size and composition. Pegmatite deposits are found in igneous rocks throughout the world. They may range up to thousands of feet in length and hundreds of feet in thickness. A great variety of valuable minerals and gemstones are mined from pegmatites: mica, tourmaline, quartz, topaz, beryl, and many others.

Beryl is used primarily as a source of the metal beryllium but occasionally spectacular crystals of this mineral are found, and they may be tinted in a number of different hues — violet, gold, pink, and many shades of green. The clear blue-green specimens, known as aquamarines, are among the most highly prized. Because these are frequently formed in pegmatites very large single crystals can be found. One aquamarine discovered in Minas Gerais in Brazil weighed over 200 pounds and produced 200,000 carats of jewels. The limpid aqua coloring of this beautiful gemstone is caused by small amounts of iron incorporated into the crystal structure.

Emeralds are also a form of beryl, but in this case chromium is responsible for the color — a deep, vibrant green. Emeralds are not found in pegmatites; they are created by a different igneous process in which stone is subjected to heating by volcanic activity and then recrystallized. The gem-quality corundum — rubies and sapphires — forms in a similar way when rock that is rich in aluminum is heated by contact with erupting magma, and the aluminum fraction separates out in liquid drops and crystallizes as aluminum oxide. Gems that form in this way from metamorphic rock have a relatively restricted space in which to grow. So large natural crystals of rubies and emeralds are rare.

In recent years it has become possible to produce some of the precious gemstones in the laboratory, using controlled temperature and pressure conditions. A study of the methods throws light on the natural processes of crystallization. The most successful of these techniques is the process for making artificial rubies. A platinum tube ten inches high and six inches in diameter is filled with a mixture containing aluminum oxide, a trace of chromium oxide for color, and several other compounds that act as dissolving agents. The container and contents are heated to 2400° F, at which temperature the solids are all dissolved. The fluid is stirred continuously by rotating the tube first clockwise and then counterclockwise. The temperature is slowly and steadily reduced about 170° F every day. When it drops just below 2260° F, ruby crystals begin to form, and as the temperature continues to fall the entire mass solidifies. Then an acid solution is used to dissolve the matrix, freeing the ruby crystals. They are superior in color, clarity, and size to almost all natural rubies. Emeralds of fine quality have also been synthesized successfully, but very large specimens have not been produced in the laboratory.

❧

More than 3,500 years ago emeralds were mined in Egypt, northeast of Aswan near the Red Sea. These "Cleopatra Mines" are now exhausted but hundreds of shafts are still visible, descending as far as 800 feet into the earth, and ancient tools dating back to 1650 B.C. have been found there.

A source of exceptionally beautiful emeralds was discovered in Colombia long before the Spanish conquest. It is still yielding the finest emeralds in the world even though many other deposits are being worked in places such as Russia, India, and South Africa. Emeralds are one of the few precious stones that are not found in Sri Lanka. The other important exception is diamonds.

The diamonds that adorned the potentates of antiquity came exclusively from river gravels in Borneo and India, especially the Golconda district near Hyderabad and a region slightly farther north between the Mahanadi and Godavari rivers, which flow down from the Deccan plateau — an enormous ancient lava formation — into the Bay of Bengal. Some of the most famous diamonds of history came from these mines, including the Orloff Diamond of Catherine the Great and the ill-fated Hope Diamond. The Kohinoor Diamond, which now graces the crown of the British queen, was owned by a Mogul emperor in 1304.

At about that same time — in 1295 — Marco Polo returned to Venice from extensive travels in Asia and amazed medieval Italy with his stories of the treasures of the East. One of his most extraordinary tales concerns a method of finding diamonds in India:

In the mountains of this kingdom it is that diamonds are found. During the rainy season the water descends in violent torrents amongst the rocks and caverns, and when these have subsided the people go to search for diamonds in the beds of the rivers, where they find many. Messer Marco was told that in the summer, when the heat is excessive and there is no rain, they ascend the mountains with great fatigue, as well as with considerable danger from the number of snakes with which they are infested. Near the summit, it is said, there are deep valleys, full of caverns and surrounded by precipices, amongst which the diamonds are found; and here many eagles and white storks, attracted by the snakes on which they feed, are accustomed to make their nests. The persons who are in quest of the diamonds take their stand near the mouths of the caverns, and from thence cast down several pieces of flesh, which the eagles and storks pursue into the valley, and carry off with them to the tops of the rocks. The men immediately ascend, drive the birds away, and recovering the pieces of meat, frequently find diamonds sticking to them. Should the eagles have had time to devour the flesh, they watch the place of their roosting at night, and in the morning find the stones amongst the dung and filth that drops from them. But you must not suppose that the good diamonds come among Christians, for they are carried to the Great Khan, and to the Kings and chiefs of that country.

This account parallels very closely a description attributed to Aristotle, who lived a thousand years earlier. We wonder how much of the story is founded on historical fact and how much is romantic fabrication. Some of the details do stretch credulity, but others correspond quite well with the principles of placer mining as practiced throughout the world from earliest recorded history.

～✦

The center of diamond mining moved from India to South America in 1728. In that year slaves panning for gold at the headwaters of the Rio Jequitinhonha at Minas Gerais in Brazil found some shiny translucent pebbles and brought them back to their masters. When these proved to be diamonds, many placer mines were opened up along the Rio Jequitinhonha and the Rio São Francisco in Bahía. Fortunes were made there and even the slaves benefited. A worker who discovered a fine stone was traditionally rewarded by a new suit of clothes and freedom. A woman slave found the 254-carat diamond that later became the Star of the South. For this discovery she was granted her freedom and pensioned for life.

But the excitement created by these Brazilian deposits was overshadowed by an even more spectacular strike in Africa. One day in 1866 a fifteen-year-old boy named Erasmus Jacobs picked up a pretty pebble on his father's farm on the banks of the Orange River in Transvaal. He brought it home and presented it to his sister to use in the game of "five stones." The father recognized that the pebble might have value and made some inquiries. The stone proved to be a 22-carat diamond, which was purchased by the governor of the colony. Appropriately named "Eureka," it was displayed in Paris and triggered an enormous influx of fortune hunters to South Africa. Using primitive panning methods, prospectors found many valuable stones embedded in the river gravels along the Orange and Vaal rivers, which drain the high plateau regions near the southern end of the African continent.

On this plateau, southwest of Transvaal but still far from the sea, another lucky find in 1870 led to the discovery of rich veins of diamond-bearing minerals extending deep down into the earth — a mother lode of diamonds! A servant boy assigned to digging as punishment for some now-forgotten crime uncovered a deposit of diamonds on the site that was later named Kimberley. Here diamonds are embedded in an unusual volcanic rock, kimberlite, that occasionally fills the necks and pipes of extinct volcanoes. The deepest mines go down about 3,500 feet but the diamond-bearing rock extends even below this.

Diamonds are a form of pure carbon, chemically the simplest of the major gemstones. In order to assume this type of crystalline structure the carbon must be subjected to very high pressure like that which is known to exist in the earth's mantle. The kimberlite in which they are embedded is composed of minerals — principally olivine, pyroxene, and garnet — that also form under high-pressure conditions and have a chemical composition characteristic of the mantle. So the rich veins of diamond-bearing ore are believed to have been created when mantle material was forced up into narrow fissures under tremendous pressure during violent volcanic eruptions.

When kimberlite is exposed to conditions at the earth's surface, it becomes crumbly and turns yellow. In fact, the first pipe diamonds were discovered in this soft yellow ground. Then as miners dug down deeper the yellow soil gave way to a harder blue-green rock, and at first they assumed that this change marked the bottom of the diamond matrix. But more gems were soon found in the blue ground. When this earth was cut loose and hauled to the surface, it changed color and began to decompose, allowing the diamonds to be separated more easily from the rock.

Making practical use of this fact, hard-rock diamond miners have developed the technique of extracting the blue ground, spreading it out in large open spaces, and exposing it for weeks or months to the rain and the sun. The disintegrated ore is brought back in trucks and fed into washing pans where diamonds are sorted from the matrix. Gigantic mechanical devices are used to facilitate moving the earth, crushing the pieces that have resisted decay, and separating the diamonds from the rocks. Enormous quantities of earth must be handled because even in this kimberlite formation, the most concentrated deposit that has been found in the world, diamonds represent only one part in fourteen million.

It is a surprising fact that the exploitation of a mother lode does not always represent the best and most economical way of mining gemstones, as the great diamond companies have discovered in the past half-century. In placer mining, nature has already separated the gems from their hard matrix. It has washed and tumbled them, sorted them into fractions of similar size and shape, and deposited them in layers that are relatively accessible. Diamond mines along the southwest coast of Africa have been taking advantage of these natural processes.

The Orange River — on whose banks the first African diamond was found — is the longest river in South Africa. It flows through the heart of diamond-bearing kimberlite country, carrying its burden of sediment down across the desolate scrub brush country of Bushmanland, across the bleak sand dunes of the Namibian coast to the Atlantic Ocean. North and south of the mouth of the Orange River this forbidding shoreline extends for more than a thousand miles. Known as the "skeleton coast" and feared by sailors for many centuries, it is a desert swept by constant southwest winds off the Atlantic, bathed by the cold, north-flowing Benguela current. Along this coast the Orange River has scattered showers of diamonds like a prodigal sower casting precious seeds upon the winds. Eon after eon the sands beside the sea have been enriched with gems. Some lie on the surface where they can be picked up by hand. Some nestle in nooks and crannies of the rocks. Many more are buried deep in the dunes. Washed out to sea, they have been returned by the prevailing winds and currents and laid up in terraces along the beaches.

These diamonds are now being collected in a tremendous mining operation that utilizes squadrons of the most powerful modern earth-moving and sorting machines. Bulldozers and scrapers peel the sand from the desert dunes, exposing the diamond-bearing gravel layers of the marine terraces, while other machines sift the loads of sand for scattered gems. Bulldozers build barricades of sand dunes, forcing back the sea and widening the beaches so the nearshore ocean bottom can be mined.

This great mechanical "panning" operation goes on six days a week, twenty-four hours a day, at De Beers Consolidated Diamond Mines north of the Orange River. Nearly sixty million tons of soil are moved annually. The diamond yield represents only one part in 200 million, seemingly a much less favorable proportion than that of the kimberlite mines. But the earth in this location is much easier to work and the diamonds obtained here are more valuable because they have already been selected by natural processes. Only those that survived the long trip downriver to the sea have been deposited along this shore. Their size, shape, and quality are relatively consistent, while the product of the kimberlite mines is a mixed assortment with many fragments useful only for industrial purposes. Most of the diamonds that have made the long journey to the Namibian shore can be cut and polished to make jewels.

Gem cutting is a fine art, involving an understanding of the nature of the stone: how its tiny identical units are built up, how the angle and spacing of the facets will affect the fire and brilliance of the finished jewel. The expert lapidary turns the rounded, translucent gem in his hand, studying it from all angles, observing its planes of symmetry, its characteristic grain and cleavage. He notes the presence of flaws and inclusions, which usually detract from its value. But he also knows that occasionally they enhance the worth if the stone is properly cut to take advantage of their presence, as in the case of the cat's eye or the star ruby. The creation of a jewel involves an unusual partnership of man and nature. It is one of the rare occasions when they work together to make something more beautiful than either could have produced alone.

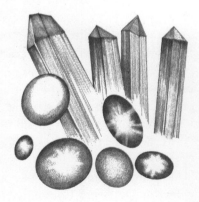

Diamonds, emeralds, rubies, sapphires, and a host of lesser-known precious stones have been fired in the crucible of volcanic action, tossed off like sparks from that glowing forge where the earth's crust is constantly reworked, new mountains are constructed, new arrangements of land and sea are hammered out. In this fiery furnace another remarkable process has also occurred — the isolation and concentration of some of the rarest and most valuable metals: gold, silver, copper, platinum, and many others.

Gold is very widely and diffusely distributed around the planet. Although it is present in all rocks, even coal and lignite, it represents only one part in 250 million of the continental crust. It is slightly more abundant in oceanic crust and even in seawater, but still the percentages are extremely small. Rich veins of gold, however, are often found in the eroded roots of old volcanoes. Deeply embedded in igneous formations, threads of this precious metal, sometimes only an inch or two thick, trace winding paths downward toward a point far below the ancient crater.

The process by which these deposits are laid down is not completely understood. But according to the most generally accepted theory, gold is carried in hydrothermal solutions — hot, salty waters containing dissolved minerals. These percolate through the cracks and fissures left by the solidification of volcanic rocks. When temperature and pressure levels are just right, gold begins to crystallize out of these concentrated solutions. Narrow passageways in the rock are the most favorable for deposition because they offer the maximum surface area, encouraging crystallization. If openings are large and continuous, the fluids may pass through without depositing much ore.

Other metals, like silver, platinum, and copper and certain metallic compounds — iron pyrites, galena, and chalcocite — are also transported in the same hot fluids and precipitate out under similar conditions. These are often present in or near veins of gold.

With the passage of time the rocks that enclose the metal deposits are gradually weathered and softened by rain and wind. As in the case of gems, flakes of gold loosened from the rocks are swept along with the other sediment and carried downstream. Since the gold is seven times as heavy as granite chips of the same size, it falls more rapidly to the riverbed and is not transported as far from the source. Some gold dust is always present in the beds of rivers that flow through regions of igneous rock formation. But the amount varies enormously and serves as a clue to the location of important deposits. For reasons that are still mysterious gold is found most often in very old volcanic formations or quite recent ones, while those of intermediate age rarely contain veins of this precious metal.

∾

Gold was one of the first treasures of the earth to attract the admiration of man. Dense and brilliantly colored, it is extremely durable, yet easy to work, and its beautiful luster is not dimmed by time. Although a rare element, it was known and highly prized even by the earliest civilizations. Wall paintings and carvings made in Egypt in 2900 B.C. depict the washing and panning of gold. Exquisitely wrought gold ornaments from ancient China, Persia, Assyria, and Crete have survived, as perfect today as when they were fashioned thousands of years ago.

In all ages gold has been a symbol of wealth and power. Wars have been fought over it and myths woven around it. It is said that the legend of the Golden Fleece dates back to an expedition launched in 1200 B.C. to seize gold that had been washed from river sands in Armenia by peasants using sheepskins to catch the flakes of gold. In that same region the river Pactolus, flowing through Asia Minor, was the source of the riches of Croesus.

The dream of turning less valuable elements into gold inspired the fairy tale of King Midas and the search for the philosopher's stone. It caused the alchemists of the Middle Ages to explore the properties of metals and opened the door beyond which lay all the discoveries of modern chemistry.

The exploration of America was also sparked by the search for gold. No one knows how early the rich deposits of this metal were discovered in South America and Mexico because no records exist of the early Inca civilization, and many of the Mayan and Aztec records were lost or destroyed. Most of the beautiful ornaments they fashioned were melted down by the Spanish conquerors. (The Spaniards promised to ransom the last Inca, Atahualpa, for a roomful of gold objects, but they executed him after taking possession of the treasure.) Under their rule the South American mines, worked by Indian slaves, continued to yield vast quantities of the precious metal. In the sixteenth century 80 percent of the world's gold production came from these deposits. Then, several hundred years later, gold strikes in North America initiated the era of the great gold rushes and the opening up of the American West.

The discovery of the fabulous Comstock lode on the eastern slope of the Sierra Nevadas illustrates a familiar pattern — a trail of gold dust led the prospectors from river sands to veins hidden in the heart of the volcanic mountains. The following account is taken from a monograph written by Eliot Lord in 1883:

> The first wagon train which entered the valley in the following spring was a noteworthy little caravan. The party were nearly all Mormons, led by Thomas Orr. . . . On the 15th of May they halted for a few hours, at noon, beside a little creek flowing down from the range of hills which bounded the valley on the east. The cattle were turned loose to graze among the sage brush, and the women of the party prepared the simple dinner of bacon and potatoes. William Prouse, a young Mormon, meanwhile picked up a tin milk-pan, and going down to the edge of the creek began washing the surface dirt. After a few minutes he returned and showed his companions a few glittering specks on the bottom of the pan. The specks were gold dust, worth intrinsically only a few cents and thrown carelessly aside a few moments later. But . . . this pinch of dust was positive evidence of the existence of gold in the deserts of Western Utah [which later became Nevada], and that starting point once given, the exploration and development of the mineral resources of the land were assured. . . .
>
> John Orr, the son of the Mormon leader, and several others returned to Gold Creek or Cañon, as they named the ravine where the first signs of gold were found, and resumed prospecting. Orr, with one companion, Nicholas Kelly, worked up the cañon rapidly until, on the first day of June, they reached a point where the banks of the rocky ravine approached so near each other that a narrow passage only was left. Through this cleft the water of the creek flowed swiftly, falling over the rocks in tiny cascades. In a crevice at the edge of one of these little falls Orr thrust a butcher's knife and pried off a loose fragment of rock. The running water soon washed out the underlying dirt, and he saw a small golden nugget which the rock had covered. In a moment he held in his hand the first piece of metalliferous quartz from a district which has yielded the great bonanza of the present century. Other bits of gold-bearing quartz and gold dust as well were afterwards obtained,

but the prospectors lacked tools and provisions, and, bent on reaching California, abandoned the cañon and crossed the Sierras as soon as the trail was sufficiently free from snow.

But meanwhile other emigrants had entered the valley, and the new placer diggings were fairly opened. The news of the discovery of gold spreads from prospector to prospector precisely as the discovery of carrion is announced by the flight of vultures and crows. A horseman riding in haste toward some point off the beaten route is an intelligible sign and lodestone. . . .

At the time when the Gold Hill placer was beginning to yield rich returns (May 1859) the miners in the cañon to the north had moved up the ravine until two of their number, Patrick McLaughlin and Peter O'Riley, were washing dirt on the slope of the hill at its head. The earth was so poor that they were fast becoming discouraged and talked of setting out for the diggings on the Walker River, which were then thought to be promising well, but they worked on, hoping to earn enough to purchase the outfit necessary for their proposed venture, and made their trench each day farther up the hill.

A little rill of water flowed down the slope toward them, but so slender was the stream that, in order to use it for their rockers,* they were obliged to dig a hole in the ground as a reservoir when they had crossed the narrow basin at the head of the cañon and reached the steep side of the Sun Peak. Accordingly, on the 8th of June they began to dig this waterhole a short distance above the cut where they were mining. The earth thrown carelessly by their shovels was a yellowish sand mixed with bits of quartz and friable black rock, such as they had never seen before. Simply as an experiment they concluded to wash a little of the sand in their rocker, though they had slight hope that it would yield any return. From the trench sand they had been getting only a few specks of gold dust, but when the new earth had been shaken in the rocker and the muddy water had drained away they saw the bottom of the cradle covered with glittering dust. Overjoyed at their good fortune they began to wash hastily the precious earth, and again and again swept up a golden layer. . . .

The line of a great lode had been cut and a modern box of Pandora fairly opened. As usual, however, a loud-spoken trickster contrived to rob the true discoverers of whatever honor is due them. The names of John Orr, Peter O'Riley, and Patrick McLaughlin are almost unknown, — the Comstock lode is famous throughout the world.

* The rocker, or cradle, is a device for sifting sediment. It handles a larger quantity of earth than the pan or sieve.

Henry Comstock . . . was a tall, gaunt Canadian, who had wandered about for many years as a fur-trader and trapper, and at length drifted to Gold Cañon, where he prospected with indifferent success. . . . By chance [he] came riding over the hills in the early evening of the day on which the discovery was made. The two Irishmen were preparing to leave the ground for the night and were cleaning up their rocker for the last time. They had marked in some way by stones or stakes the limits of their claim, and had no reason to suppose that their rights would be disputed. As he [Comstock] approached the water-hole his restless eye caught sight of the earth heap and the rich contents of the rocker. Without a word he sprang from his pony and threw himself on his knees in the hole, running his fingers through the dirt and observing the tiny spangles which clung to his hands. Standing up at length he coolly informed the astonished miners that they were trespassing on his land. A goat might search in vain for food among the crags and crumbling rocks of that barren peak, but Comstock pretended that he had taken up a tract of 160 acres along the slope of the mountain for a ranch, and that the conveniently indefinite boundaries of this tract included the water-hole and trench where the men were working. He was willing, however, to allow them to continue work, if they in turn would recognize his proprietary rights by conceding to his friend Emanuel Penrod and himself an equal share in their claim.

A more preposterous demand could scarcely be made. He had no title to the ground and no record of his ranch location. The land in question was the property of the United States, open by tacit consent for the occupation and use of any citizen. Yet he talked so loud and long that the miners for the sake of peace agreed to his proposal.

O'Riley and McLaughlin had penetrated one of the veins of gold and silver that ran in a complex network throughout the mountain beneath their feet. But they were not made rich by this discovery. Their rights were usurped not only by Comstock but by many other unscrupulous fortune-hunters who descended on the scene. In the end eighty-six companies shared the treasure of the Comstock lode, which produced gold and silver worth more than $378,000,000.

The structure of the Comstock lode is typical of most gold and silver deposits *in situ*. Many slender veins of metal lead in winding paths down toward the center of the mountain where they join together in a single vein that descends to unplumbed depths like an umbilical cord leading down toward the soft interior of the earth. (Perhaps it is no coincidence that a great volcano like Mount Agung has been called the Navel of the Earth.)

In this deposit, as in most other mines, gold is found in the native state, not in chemical combination with other elements. Gold is the least chemically active of all the metals. In the native state, however, it frequently occurs along with silver, platinum, palladium, and copper; so various alloys of these metals are common products of gold mining. They may be used as alloys or refined to recover the pure gold. When the Comstock lode was discovered, hydrothermal solutions like those from which the ores had been deposited were still seething in the deep interior of the mountain. Extraordinarily high temperatures were encountered in the mines, and great volumes of scalding water poured out of many fissures.

Although gold is usually found in the native state, some important deposits of gold-bearing minerals have also been laid down. At Cripple Creek, Colorado, for example, gold was deposited as tellurides like sylvanite and calverite, which are chemical compounds of gold, silver, and tellurium. The Cripple Creek mining area is enclosed in a 10,000-acre bowl of volcanic rock with impermeable granite walls extending upward to altitudes of 9,200 feet. Ground water that collected in the bowl had no outlet, and miners encountered water in their shafts at depths of about 800 feet. Pumps had to be used to keep them dry until a tunnel was bored through the granite in 1904.

One day miners working deep in the labyrinth of shafts at Cripple Creek thrust a flare through a large hole in the rock wall.

What the three men saw stunned them as a child is stunned by his first Christmas tree. It was a cave of sparkling jewels. The brightness blinded them at first but then they made out that the jewels were millions of gold crystals — sylvanite and calverite. Spattered everywhere among the crystals were glowing flakes of pure gold as big as thumbnails. The cave was forty feet high, twenty feet long and fifteen feet wide. Small boulders glittered on the rough floor. Piles of white quartz sand glowed like spun glass. This was an Arabian Nights scene in the twentieth century. . . . The cave was called a "vug." Technically it was a geode — a hollow, rounded nodule of rock lined with gold crystals.

Here in the heart of this geode we are brought face to face with a question that has been posed many times but never satisfactorily answered. By what magic does nature draw together the minute bits of precious elements that have been widely dispersed throughout the rocks and waters of the earth? How does it segregate these and lay them up in the remarkably pure form in which they are found? As the hot, rich solutions containing atoms of many elements percolate slowly through the rocks, temperatures and pressures gradually drop. Then suddenly, as though touched by the philosopher's stone, atoms of one kind leap from the formless liquid state and, joining together, build that elaborately designed construction in space — the crystal. The presence of just a single crystal or particle of dust or small protrusion along the rock wall can start the transformation of formlessness to form. Atoms of gold — or silver or carbon — line up in precisely ordered patterns like dancers taking their places in a quadrille when the band strikes up. But the scale and speed of the action surpass any examples in our human experience. The growing of a typical crystal requires the proper placement of something like sixteen trillion atoms an hour.

It is this act of crystallization that has concentrated and laid down rich hoards of treasure in the earth. In fact, the entire lithosphere is an intricately interwoven fabric of many crystals: rocks and sand, gold and tin, diamonds and ice. Crystallization is perhaps the most important single process that creates the world we know. It is an expression of the supremely logical structure underlying all things — an orderliness usually hidden from our sight beneath the ever-changing masses of clouds and soil and sea. As Loren Eiseley said, "It is an apparition from that mysterious shadow world beyond nature, that final world which contains — if anything contains — the explanation of men and catfish and green leaves."

12

Tides of Life

> Where air might wash and long leaves cover me;
> Where tides of grass break into foam of flowers,
> Or where the wind's feet shine along the sea.
>
> — A. C. SWINBURNE, "Laus Veneris"

In the Bristol Channel on the coast of Wales, as on many other beaches around the world, the land melts very gently into the sea and the receding tide slips silently down long stretches of shining shore, baring sand ripples, leaving tiny stranded pools of water like dimples on the beach's face, and exposing deep wrinkles that mark the course of the water's retreat.

The tide is out. The moon has passed nearest to this place in its diurnal path around the earth. She has piled up the waters of the planet as though she would gather them unto herself, drawing them across the great distances of empty space to wrap her parched, sterile, sun-burned body in this cool, life-giving fluid. But day after day she fails to accomplish her purpose. Her hold slips on the waters of the earth as she sails inexorably on around her orbit. Very slowly, almost im-perceptibly, the waters begin to flow back again across the beaches.

A flood of living things has ebbed and flowed across the planet like the tide. Drawn from some deep and unfathomable reservoir, life made its appearance on earth at a time that cannot be precisely marked. From imperceptible beginnings it grew and gathered strength as it spread in a green flood across the planet. It altered the color and shape, the fragrance and texture of the land. Countless trillions of plant roots, the burrows of worms, and the pathways of microscopic creatures aerated the crust, making soft friable soil out of bare sand and hard rock. Drifts of spring flowers and autumn leaves filled the valleys with scarlet and gold. Even the winds were tinged with the breath of life. Sweet and pungent smells enveloped the earth — the odor of decaying leaves, the fragrance of honeysuckle and ripe clover.

But this living tide has receded, too, not once but many times. Its rising and falling seems to have followed a periodic movement although there is no clear, predictable cycle like the ocean tides.

Just a few years ago scientists thought they knew how life began. It was created by pure blind chance as the essential ingredients were shuffled and rearranged throughout many eons. The theory, repeated so many times, took on the character of accepted fact. Once upon a time — according to this tale — the earth was sterile as the surface of the moon. The naked, newborn rocks of the continents were bathed in deadly radiation from the sun. The oceans were warmed and in their depths elements were slowly agitated, meeting and parting in a haphazard dance that continued for perhaps a billion years. Given this great length of time, it was inevitable that the right molecules would meet by chance and form the complex organic molecules that were the necessary precursors of life. Gradually these organic substances accumulated until the primordial seas were thick with them. Again at least a billion years passed while the continual agitation of this warm rich soup brought all possible combinations of the molecular building blocks together and finally the first living thing came accidentally into being. Once such an organism was formed, it fed on the organic compounds and multiplied, protected by the water medium from the lethal ultraviolet rays of sunlight.

According to this scenario, a long, lifeless period in the existence of our planet was a necessary condition for the origin of life on earth. "There seems a grim inevitability about the scheme," wrote Isaac Asimov in 1960. "Given a planet with the proper kind of chemistry, with a temperature that is neither too high nor too low, with an adequate air supply made up of the right gases, with an ocean, and with a sun of the right type shining down — and, most of all, given enough time! — it would seem that with grand relentlessness, first nucleic acid molecules would form, then cells, then chlorophyll, then multicellular creatures." Such was the accepted dogma just a decade or two ago, but now the latest scientific information indicates that the essential ingredient — long periods of time — never existed. As geologists probe further and further back into the early history of the planet, analyzing the most ancient rocks, the advent of the first living thing has been pushed back into the early infant years of the planet.

Geologists believe that the earth is 4.6 billion years old, but for many millions of years after its birth the planet was too hot to permit crystallization of a lithosphere. The oldest known rocks of the crust are in western Greenland. They date from 3.8 billion B.P. and probably represent a portion of the first permanent crust of the planet. Even here — as far back as science has been able to probe into earth's history — evidence or at least hints of life are present. Organic molecules characteristic of living organisms have been discovered in this formation. They occur in metamorphic rocks which are usually devoid of signs of life because they have been subjected to heating and pressure from igneous activity. But in these rocks several small pockets seem to have escaped melting and recrystallization and some organic molecules survived. They are similar to carbon compounds produced by biologic activities like photosynthesis. This finding suggests that the compounds may have been formed by living things.

Definitive proof of the existence of living organisms has been found in rocks 3.5 billion years old, organisms that were already advanced beyond the most elementary phase. Thus the time in which the building blocks might have evolved by chance has been reduced from a billion years to a vanishingly small slice of geologic time. Perhaps the building blocks were formed elsewhere. Perhaps they were synthesized in the intense heat of lightning flashes during the turbulent early years when the earth's crust was forming. Perhaps they were already present on the newly formed planet — a final gift from the solar nebula. Or perhaps they were brought to earth by comets and shooting stars. These suggestions would have met with ridicule a few decades ago, but now they are receiving serious scientific attention.

Although the oldest crustal rocks yet discovered are 3.8 billion years old, even more ancient rocks have been found on the earth. These are pieces of stray matter from the solar system that were attracted by the earth's gravitational field. More than a million billion of them enter our atmosphere every day. Usually they burn up long before they reach the surface, leaving nothing but ashes and dust to float slowly earthward. Several tons of this dust enter the atmosphere each day, but only the largest rocks arrive at the surface with a portion of their substance still intact after their flaming descent through the atmosphere. Such meteorites are samples of worlds far beyond our planet. Almost all of them are very old: the time of their formation ranges around 4.6 billion years ago, the birthdate of the solar system.

Meteorites vary considerably in composition. Some are metallic, others are rocky with a wide range of chemical constituents. Because these pieces of matter were not subjected to erosion from rain, wind, and flowing water as the earth has been, they are thought to have changed much less since their formation than the earth itself has. One type of stony meteorite contains an almost complete assortment of the elements that make up the solar system as a whole and these elements are present in approximately the proper proportions. The meteorites in this particular group, known as Type 1 carbonaceous chondrites, are believed to be relatively unaltered samples of the material that condensed from the solar nebula during the last stage of its cooling.

Imagine then the great excitement that was generated among geochemists in the 1950s when carbon compounds characteristic of life — even the important amino acids — were identified in these meteorites! Several experts also found strangely shaped inclusions they believed to be tiny fossil cells. Unfortunately, it was subsequently shown that the unusual cell-like formations were contaminants introduced in the laboratory. (Some of them appeared to be fragments of Kleenex!) This embarrassing episode pointed up the need for absolutely sterile conditions in analyzing the meteorites, and for a number of years all reports of interesting findings in these bits of extraterrestrial matter were greeted with skepticism. But today, after several decades of perfecting the techniques, it has been demonstrated beyond any reasonable doubt that large organic molecules are present in many freshly fallen carbonaceous chondrites. The building blocks of life appear to have been formed as early — or at least nearly as early — as the solar system itself.

In the 1960s another surprising discovery was made. Organic molecules were found to exist in cosmic dust clouds far out in space beyond the solar system. Their presence in interstellar space ran counter to accepted scientific theories. The chances were thought to be impossibly small for the proper atoms to encounter each other and form complex molecules in outer space where atoms are very widely separated, not densely packed as in the hypothetical "primordial soup." Furthermore, the fragile organic assemblages were believed to be too delicate to survive in the ultraviolet radiation that pervades the space between the stars. But there they are, nevertheless, molecules of methane, ether, wood alcohol, formaldehyde — the list keeps growing. Tumbling end over end in space, they send out their own distinctive radio signals, which have been picked up and identified here on earth.

Experiments performed in the late 1960s suggested a new theory about the formation of the organic precursors of life. Common clay may have acted as a mold or template for the essential proteins. The molecular structure of the clay consists of stacked sheets of silicate separated by layers of water molecules. This arrangement provides an enormous amount of surface area. Certain large molecules that form from amino acids and phosphates are held on this surface by electrical forces between the molecules. As a result long chains consisting of fifty or more amino acids are formed on the clay surface, producing the protein building blocks common to all living things. Biologists are quite excited about this theory, which suggests that the matrix of life has been widely distributed throughout the earth's crust since the lithosphere first took shape and, in fact, may have been present in many places throughout the solar system. Clay is derived from the combination of feldspar and water, substances that are ubiquitous throughout the earth's crust, in the rocks of the moon, and even in the carbonaceous chondrites.

It is beginning to look as though the oft-repeated story of how life evolved in the primordial soup is just a fairy tale. No one knows yet where the new thoughts will lead or how all the questions will be resolved. The beginnings of life are still hidden from us and no one can point with certainty to the moment when the flood tide of self-replicating molecules began.

❧

The most ancient fossils of cellular organisms have been found in rocks 3.5 billion years old in the Warrawoona Formation in western Australia. The advent of the cell with its protective membrane was a very advantageous development in the evolution of life. This membrane transmits molecules like sugar and amino acid, but larger molecules like starch and protein cannot pass through. They are trapped within the cell. Thus the organism can absorb food particles, build them up into large molecules, and accumulate them for later use. The invention of the membrane gave cellular organisms an advantage over the simple self-replicating molecules. Eventually they supplanted most of the more primitive forms.

Another very important evolutionary step had taken place by the time the Warrawoona Formation was laid down. Chlorophyll had made its appearance — the molecule that absorbs sunlight and stores it in organic compounds to be used as a food supply by the living cell. The blue-green algae whose imprint is preserved in these rocks were photosynthetic organisms. They grew in colonies, making thick mats of vegetation that were arranged in layers like a head of cabbage.

As complexity continued to evolve, eucaryotic cells made their appearance. Each of these has an organized nucleus containing paired strands of genetic material. And it possesses organelles — tiny units that perform specialized functions. For example, mitochondria facilitate the processing of food. In the eucaryotic cells chlorophyll is concentrated in organelles called chloroplasts. These advances in cellular structure improved the efficiency of the organism and made sexual reproduction possible. In the Bungle-Bungle Formation of Australia a layer of dolomite one and a half billion years old contains fossils that may be the remains of a sexually reproducing living cell. Organisms of this degree of complexity are definitely known to have existed one billion years ago, and photosynthesis was occurring widely throughout the seas.

How did these more complex life forms evolve from simple cellular organisms? Here again, a new scientific insight has revolutionized our view of the development of life. Instead of the older theory which postulated that increased complexity occurred by mutations within an individual organism, biologists now hypothesize that these more highly articulated cells were formed by the union of several different types of single cells, each one giving up its separate identity to become an essential part of the larger whole. Mitochondria, for example, specialize in processing food, while chloroplasts capture and package the energy of sunlight. Both the chloroplasts and mitochondria have their own DNA and reproduce themselves. They are transmitted as whole organelles within the egg and they closely resemble certain types of bacteria. It is theorized that they may once have been free-living bacteria that joined with other primitive cellular organisms. By combining their resources they created a more effective living unit. This interesting idea has implications that reach far beyond cellular biology. The relationship of the individual to the group takes on new dimensions if it is shown that nature builds more advanced forms of life by submerging the individual in new and more highly organized states of matter.

These earliest steps in the evolution of life took place in the sea. Thirty feet or more below the surface organisms were well shielded from the lethal ultraviolet rays of sunlight; so the oceans provided a soft, yielding, protective medium for the sensitive living creatures that inhabited its depths.

Ocean environments are influenced by many things: by latitude, by land and sea relationships, by ocean currents. Based on principles derived from the present distribution of various types of aquatic life and the record left by the fossilized shells and molds of untold numbers of sea creatures, it is possible to reconstruct the environments of the ancient seas.

Warm waters like those in the tropics today provide a very favorable situation with a large number of microhabitats. Such warm seas contain the greatest variety of life forms. Diversity tends to decrease with distance from the equator, and this fact is sometimes used as a rough indicator of latitude at various times in the past. Near the equator the warm surface waters are stirred and overturned by currents caused by the spinning of the earth. These currents bring deep water up to the surface and provide a constantly renewed source of nutrients. So these low latitudes represent an exceptionally favorable environment, and large numbers of diverse aquatic life forms flourish there. The position of the equator is actually marked for at least 5,000 miles along the bottom of the Pacific from the international date line to the Galapagos Islands with a layer of chalk several hundred miles wide and half a mile deep at its thickest point. The chalk has been formed from skeletons of billions upon billions of sea creatures that lived and died in these tropical waters. Ancient chalk layers now buried deep beneath deposits of clay have been found in cores taken by the Deep Sea Drilling

Project in the Pacific Ocean. They delineate the position of the equator at various times in the past as the plate carrying the Pacific Ocean moved northward.

The presence of an ancient tropical seaway is recorded in a vast expanse of limestone rocks extending from southern Europe and North Africa all the way east to Indonesia and west through the Caribbean, the Gulf of Mexico, and Central America. This sea, called Tethys — mother of oceans — contained an extremely diverse and distinctive assemblage of corals, sponges, ammonoids, and other mollusks. About 150 million years ago the Tethyan fauna was remarkably homogeneous, indicating freedom of movement within this domain. Later, the North American and Eurasian fauna became more divergent. This change seems to document the opening of the Atlantic Ocean and the disruption of the western part of the Tethys Sea.

The eastern portion of the Tethys is now the Malay Archipelago, and it harbors the richest community of marine fauna in the world. The aquatic life of the northwest Indian Ocean and the islands of the central Pacific is derived from this Malayan center, but with increasing distance the number of species decreases. In general, dominant species come from favorable centers while the more primitive forms are often forced to the fringes of a well-populated, successful province. For example, it has been observed that regions of increasingly higher altitude or greater depths are populated by more and more ancient and primitive creatures. Many of the species that become extinct are the ones that have been pushed out to the far edges of the region they formerly occupied. Finally, they are expelled from these last refuges by more advanced species migrating from the favorable center.

In a rigorous, rapidly changing environment like the Arctic the biological communities usually consist of a small number of species with high reproductive rates. When adequate food is available, as, for example, where waters are rich in nutrients, extreme environments can support immense numbers of individuals.

No matter how difficult the environment, life has found a way to colonize it. There is no place on earth where life has not staked out a claim — even in the 700° F heat of hydrothermal vents, even beneath the glaciers of Antarctica, on the sands of the Sahara, and 17,000 feet below the surface of the earth.

The flood tides of life need no explanation. Moved by its own inner drive, life expands and spreads across the planet. But the ebb tides are not so easily explained. At least five times in geologic memory vitality has receded from the oceans and the continents, surviving only in little pockets like the tiny pools left stranded on beaches when the sea is out. These events have caused more speculation than any other occurrences in the history of the earth.

The life that inhabited the very early seas was all soft-bodied. There were no shells or skeletons to leave well-preserved fossils. The shallow sea environments of 600 million years ago contained a growing variety of jellyfish, worms, and blue-green algae. Great accumulations of algal growth provided the basis for early aquatic food chains and created undersea gardens that were widespread throughout the tropics. But the presence of these oldest forms of life is documented only by occasional impressions left in the sand and the mud. The record is so fragmentary that paleontologists have not been able to trace the evolution of these very early life forms or to draw any firm conclusions about the impact on life of the Precambrian glaciation (700–600 million B.P.) that held even the equatorial regions of the earth in its icy grip.

Many eons later — about 570 million years ago — during a period when warm, shallow seas covered much of the land, shelled aquatic life suddenly made its appearance. The earliest of these shelled fauna were mysteriously unlike any organisms that inhabited the seas in subsequent eras. In a very short time they became extinct and other types like our oysters and clams and snails evolved. The beetle-like trilobites scurried across the ocean bottom, leaving trail marks and molds of their hard, chitinous shells. By 480 million years ago the first fauna with a central nerve cord surrounded by bone was present in the sea. These primitive fish had heavy bony armor around their heads and suckers instead of movable jaws. They probably swam clumsily along the seafloor, grubbing in the mud for food.

Most of this early life was dependent upon a warm, shallow marine environment. The climate was mild and equitable. For many eons the sea level rose and epeiric seas were very widespread. Then about 440 million years ago a new ice epoch occurred. The formation of glaciers caused a great drop in sea level, draining the shallow seas. The severe shrinkage of the shallow marine habitat and the low water temperatures were so unfavorable that many species of aquatic life died out.

Eventually these glaciers, too, melted. The tide turned again and flowed back along the ancient beaches. Life spread out and rode high along with the tide. The number of marine species increased in a spectacular way. Cephalopods whose living relatives include the octopus and squid left large numbers of distinctive funnel-shaped shells. Attaining frightening proportions, often exceeding twenty feet in length, they appear to have ruled the Ordovician seas. Rugose coral developed and flourished and even smaller organisms — bryozoa and graptolites — formed widespread colonies. Rich beds of fossils laid down in these epeiric seas are found in many inland locations today.

For a period of perhaps ten million years fluctuations in sea level occurred. These oscillations may have led to one of the most important events in the history of the earth: the invasion of the land by life about 400 million years ago. Long before that time photosynthesis had caused free oxygen to accumulate in the atmosphere and an ozone layer had built up, screening out the most damaging ultraviolet rays of sunlight; so living organisms could exist without a protecting layer of water. But even after the appearance of the ozone layer there were formidable difficulties with adapting to life on the continents. Gravity was a troublesome force to be dealt with, making all movement and transport of nutrients more difficult. In the sea gravity was counteracted by the buoyancy of the liquid medium, but on land a strong supporting structure for the organism was required. Temperature extremes were much greater on land than in the ocean, and there was constant danger of desiccation. A regular supply of water must reach every cell.

The first land plants that left their imprint in the fossil record had no true roots or leaves, but they had developed a vascular system enabling them to raise water from the base to upper portions of the branches. Stiff, thorny trunks provided the supporting structure, and the cells of the plant contained chlorophyll, so the plant could manufacture its own food from air and sunlight. These characteristics had evolved in the sea. Then a sudden reduction in the extent of the shallow seas provided the opportunity for these plants to exploit their special attributes and pioneer a new environment.

Land plants diversified and evolved very rapidly. By 350 million years ago a tide of green vegetation had flowed across the lowlands, following the watercourses. By this time, animals, too, had begun their colonization of the continents. The first air-breathing fauna were spiderlike creatures dating from about 390 million years ago. Then thirty or forty million years later, a fish with bony fins crawled up onto dry land. There on the edge of a brackish and retreating sea, ringed with swampy fern forests, began the great adventure of exploration and challenge that led along devious paths to frogs and snakes and crocodiles and pterosaurs and man.

This fish — actually the first amphibian — had two special attributes that made the change in habitat possible. It had bones in its fins so it could "walk" on dry land. And it had internal nostrils that enabled it to breathe with its mouth closed. In the shallowing ponds where insufficient water was available, this amphibian could breathe air. But even with these advantages it was still tied to a water medium. Its eggs had to be laid in water to prevent their drying out, and larval stages had to be lived in the sea.

When the first amphibian left the warm cradle of the tropical waters and clambered onto a barren shore, it entered a harsh and frightening environment. The stones were sharp beneath its tender fins. The brutal weight of gravity pressed down upon its body, making every movement a painful effort. The wind chilled its wet scales and the sunlight pierced its unaccustomed eyes like a sword. If life had followed simply the path of least resistance, this amphibian would have remained in the drying pond and died there. But something caused this ancestor of ours to take the first torturous step along a new evolutionary path.

Wherever chance provides an opportunity, life forms reach out and take advantage of it, releasing the potential mysteriously locked up within the genes. In this very important way the tides of life are different from the ocean tides. The waters of the earth always flow downhill, responding passively to the forces acting on them. But living organisms travel upstream against the current. They are stimulated by adversity; they accept new challenges. They move out —sometimes by necessity, it is true — but often voluntarily from easy to difficult environments.

During the Devonian Period (395–345 million years ago) the many and various potentials of life were rapidly unfolding in the oceans as well as on land. Very extensive coral reefs were built up, forming a complex interrelated community of aquatic life and creating new habitats where many different species could find food and protection. The enormous quantity of reef structures dating from this period documents the wide extent of conditions favorable for the growth of coral: warm, shallow, sunlit seas.

In the deeper waters several marine monsters evolved. The Arthropoda produced sea scorpions that sometimes attained lengths of ten feet. The huge cephalopods of earlier time gave way to coiled ammonites. But these, too, grew very large — about ten feet in diameter. Sharks had made their appearance and an even more formidable species of jawed fish with heavy armored plates on its head and body. These placoderms were voracious predators, attacking even the sharks, and dominating the Devonian seas.

For a long period of time ocean levels were high. Then about 340 million years ago the fossil record shows a gradual change from marine to nonmarine conditions. Much more land stood above sea level. What had been a paradise for denizens of the deep became luxuriant swamp forests inhabited by giant lizards and insects. Wholesale extinctions of shallow marine organisms occurred. But at the same time many new forms of life evolved on land. Conifers and other large seed-bearing trees clothed the uplands of the great landmasses with dense green forests. The valleys were lush with ferns and mosses. Reptiles evolved from amphibians. They were more independent of the water medium because their eggs had an outer case or shell that allowed oxygen to enter but prevented liquids from escaping. These eggs contained food for the embryo and even a storage area for wastes. They were self-contained units that could be incubated on dry land.

During the next 100 million years sea levels passed through a number of minor fluctuations. When the shallow seas retreated, the land was dotted with marshes and broad plains. Then slowly but inexorably the average levels continued to drop. Like the ebb of a great neap tide, the waters reached their lowest point at the end of the Permian period 225 million years ago. The epeiric seas drained away; the continents were all emergent. And the greatest extinction of life ever to affect the earth occurred. Almost all marine species died out (95 percent) — all trilobites and sea scorpions, all ancient corals, and placoderms. Many other fish and mollusks disappeared entirely from the face of the earth.

In the history of the great extinctions from 440 to 200 million years ago, there seems to be a very striking correlation between the ebb and flow of life and the rise and fall of ocean levels. Species died out at low tide and were born at flood. Water is truly the elixir of life.

There is a folktale in England held especially dear by the people who live along the sea. Most deaths happen at low tide. "People can't die along the coast except when the tide's pretty nigh out," wrote Charles Dickens. "They can't be born unless it's pretty nigh in — not properly born, till flood."

Since ocean levels drop dramatically during a major glaciation, it is reasonable to look back over the history of the ice epochs in seeking the cause for the great extinctions of life. However, in only one case — the Ordovician, 440 million years ago — is there a close correlation in the timing of these two events. The Devonian extinction preceded the onset of the Carboniferous-Permian glaciation, and the Permian extinction took place when the earth had emerged from this ice epoch.

Several other reasons for changes in water level have been suggested. Earth movements and variations in the rate of seafloor spreading could have affected the size of ocean basins and the extent of the continental shelves, but there is no real consensus among geologists on this subject.

After the Permian Period the waters began to rise again very slowly while the large landmass — Pangaea — was beginning to split apart, forming separate continents and new oceans. About 195 million years ago an abrupt recession of ocean levels occurred and a die-off sharply reduced the number of marine species. Many clams and other mollusks became extinct. The ammonites, which had diversified into about 400 genera and families, were almost entirely wiped out. Only one family survived this crisis, but later it multiplied and radiated widely.

In the eons following this episode sea levels continued their general upward trend, reaching an all-time high water mark about 100 million years ago. Then there followed a long period of receding waters. The last major extinction occurred sixty-five million years ago. Although this most recent catastrophe in the history of life followed an extended period of decreasing ocean levels, the die-off did not occur at the lowest point of the tide. The water levels oscillated and, on the average, continued to fall, reaching their minimum about thirty million years ago.

This latest extinction affected life on land as well as life in the seas. It wiped out the dinosaurs, the large marine reptiles, the flying reptiles, the last ammonites, and most of the marine plankton. Land plants, however, were not especially affected; neither were the organisms of the shallow coastal waters. Some small land animals like the marsupials died back but others — for example, our own mammalian ancestors — throve and multiplied.

Many theories have been proposed to explain the extinction, but none of them was supported by any real evidence — until 1980, when geologists discovered very significant differences in the chemistry of rock strata sixty-five million years old. These layers contain an unusually large proportion of certain elements that are relatively rare in the earth's crust: platinum, iridium, lead, tin, copper, nickel, and zinc, to name just a few. Iridium (a first cousin to platinum) shows the most spectacular increase in abundance. The earth's crust is depleted in these minerals because they are heavy and when the planet was hot enough to melt, these heavy elements migrated toward the center under the influence of gravity. Certain meteorites — the carbonaceous chondrites, in fact — have larger proportions of these elements.

As a result of these discoveries, it has been suggested that a very large meteorite about nine miles in diameter struck the earth sixty-five million years ago. The impact dug an enormous crater, throwing up great clouds of dust enriched by iridium and the other heavy elements. This dust spread throughout the upper atmosphere and reduced the amount of sunlight reaching the earth for several years. Photosynthesis was drastically reduced, causing the death of land plants and marine plankton. In the absence of plants, the herbivorous dinosaurs died of starvation and carnivorous types starved also because the animals they preyed upon were gone. Dead and dying vegetation, washed down into coastal waters, enriched the shallow seas, providing enough food for the organisms that inhabited those waters. Small terrestrial animals survived on seeds and nuts. The seeds of land plants remain viable for years and these plants regenerated after the dust subsided.

This scenario seems to fit the facts reasonably well, although there is some evidence that the extinction of the dinosaurs did not occur quite as suddenly as implied by this theory. Further confirmation is needed before it is considered proved. If a plausible distribution pattern of the enriched elements around the globe can be demonstrated, the theory will rest on a firmer basis. The discovery of the crater left by the meteorite would be the best evidence. The impact is believed to have dug a hole at least ninety miles in diameter.

On earth, wind and rain obliterate the craters left by giant meteorites, but space exploration has shown us that these events have occurred frequently in the solar system. Many planets and moons are deeply pitted with craters. The rate at which meteorites of various sizes have impacted the earth and the moon has been calculated. On the average an object larger than one-half mile in diameter can be expected to strike the earth every 100 million years. Four of the great extinctions have occurred at intervals of about 100 to 150 million years. So here is another possible explanation of the rise and fall of life around the world.

If the meteorite theory proves to be correct, a strange irony emerges from this history of life. A rendezvous in space with scraps of primordial matter may have distributed the building blocks of life on earth billions of years ago. Then another such event brought life to an abrupt end for many species. Thus the dinosaurs and the ammonites may owe their beginning and their end to these cosmic encounters. Mankind, on the other hand, may owe its existence twice over to these flaming messengers from space. The small mammals from which we are descended had lived for almost 100 million years in the shadow of the dinosaurs without undergoing any spectacular evolutionary development. The sudden demise of these very successful reptiles provided the opportunities for mammals to evolve, radiate, and take over the dominant roles on land.

Even the great extinctions provide opportunities for life to reach out and attain new levels of achievement. Unlike the ocean tides, the rise and fall of vitality across the earth is not simply repetitive. On each flood tide rides a dazzling array of new creations: the coral reef, the dragonfly, the golden eagle, the apple tree. In a manner reminiscent of the other cycles we have traced throughout the history of the earth — the recurring ice epochs, the elevation and leveling of the great mountain ranges — the tides of life never bring back the past. From the first blue-green algae to the questing mind of man, the unfolding of potential has proceeded in a totally unique manner in the time dimension.

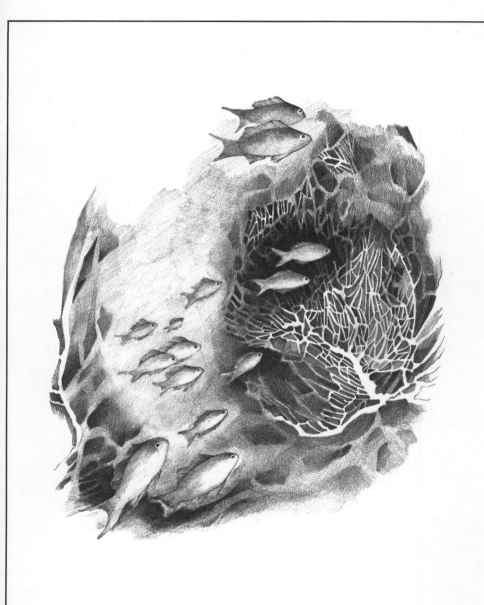

13

The Living Lithosphere

> Full fathom five thy father lies;
> Of his bones are coral made;
> Those are pearls that were his eyes:
> Nothing of him that doth fade
> But doth suffer a sea-change
> Into something rich and strange.
>
> — SHAKESPEARE, *The Tempest*,
>   Act I, Scene 2

We floated in blue — blue sky above, blue water below. The sky was smooth and uniform as an inverted cup; the ocean was a dazzling disk of many hues, shading from pale transparent aqua where sandbanks lay in shoal waters to deep regions saturated with color, ultramarine and peacock blue. Here and there mysterious patches of purple betrayed the presence of something lying just beneath the surface of the sea. Veiled glimpses of shadowy shapes came and went as small waves flecked the surface with gold. Here lay the Great Barrier Reef, which I had traveled 9,000 miles to see.

On the horizon a long low island rose just above the water level. Its dark outline seemed to breathe and flow. Occasionally the whole surface exploded in a gray cloud that rose swiftly up, fragmented against the sky, and settled slowly back again like ashes stirred by a vagrant breeze. The sooty terns were nesting on Michaelmas Cay. Thousands of them covered the tiny sandspit, turning the dazzling shore dark with their gray wings.

I looked down from our boat where it rode at anchor through water so transparent that the long slanting anchor chain was clearly visible, link by link, until it disappeared beneath the keel. Just off the starboard side I could see the outline of a large coral reef — a "bommie," as this type of formation is nicknamed in Australia. The name is short for *bombario*, the aborigine word for an isolated coral head that rises up from the ocean bottom.

Several of the other passengers on the boat were already dressing in their diving clothes. They helped each other into tight wetsuits, strapped on the compressed air tanks and weighted belts, adjusted the pressure gauges, and finally put on flapping black fins and masks with long flexible breathing hoses. Now they looked like visitors from another planet. What an appropriate costume in which to explore a strange and alien world!

The underwater world is indeed an alien place where the newest crust of the planet forms the incongruous background for some of the most ancient forms of life. Invading their world is like traveling in a time machine back to those far-off days before the first slippery sea creature climbed out onto a forbidding shore, bringing with it the encapsulated seed of the future. From that moment on, the site of explosive evolutionary growth was the land; the hidden world of the sea was left behind. In all those millennia of time the life of the ocean has continued patiently to spin out its elaborate web of birth and death, growth and decay, cooperation and competition, relying on principles invented long before the dinosaurs left heavy footprints in the swamps beneath the tall cycad trees of the Mesozoic forests. Except for a few latter-day invasions from that other world — the whales and dolphins and, more recently, man — almost everything that lives and grows beneath the surface of the sea is built on a very ancient plan, tracing its ancestry back to some of the first families that colonized the planet. And yet, in a number of very important ways, these aquatic life forms have been more successful than their counterparts on land.

❧

As the divers finished their elaborate preparation, they flopped awkwardly to the stern of the boat where the diving gate had been opened, and eased themselves off into the water. I had already put on my short orange wetsuit, adjusted my snorkel and mask, sucking in my breath to test the fit. Then I stood at the diving gate, looking out at the sea.

"Are there many sharks in this water?" I asked our guide.

"Oh, yes," he said, "several types are frequently seen here, but they won't bother you once you are over the bommie. It is too shallow for sharks there."

I knew that Mike spoke with authority because he was a professional shark hunter employed by the state government of Queensland to set nets and discourage the sharks from coming near shore. On his days off from shark hunting he chartered his boat for snorkelers and divers. Earlier that day Mike had brought out a collection of photographs he had taken of sharks. One startling picture showed the contents of a shark's stomach with an enormous sea turtle that had been swallowed whole, shell and all.

Now I felt a moment's hesitation, but I knew it was too late to turn back. I slipped off into the water and as my mask broke through the reflecting surface of the sea I entered a veritable fairyland — like Alice stepping through the looking glass.

An underwater landscape flooded with sunlight lay just a boat's length away and in a few moments I was floating over this magic garden. The water was only six or seven feet deep and very clear. The sun shone through with undiluted brilliance except for the ripples that made an ever-changing movement of light and shadow across the bommie. Many types of coral were growing there — purple and mauve gorgonian fans, daisy leather coral like tiny blossoms, gold leaf, and yellow cauliflower, and staghorn coral tipped with cobalt blue. Brightly colored fish moved in clouds among the fronds and branches. Drifts of electric blues hovered near a mound of gold leaf coral and darted back within its recesses as I approached. Beautiful iridescent angelfish moved like butterflies from flower to flower. An orange clownfish peered cautiously out between the waving arms of a pale pink sea anemone. This clever little fish manufactures its own protective coating so it can live within the circle of these poisonous tentacles without suffering any ill effects and thus be protected from all its enemies. As it rubs itself lovingly against the anemone's tentacles, a thin layer of mucus is produced, which is a chemical signal to the anemone saying: "This is me — the clownfish — don't sting me. I'm your friend. I will swim just outside your fronds and lure bigger fish in near where you can catch them easily." The gaudy color and markings of the clownfish make it a very effective decoy.

Vibrant color is a dominant trait of life on the coral reef as it is everywhere in nature. Living things are often brilliantly hued. Dead things are gray or brown or white when they have been bleached by the sun. The color of a coral reef community is a measure of its vitality; a dead reef looks like a pile of old bones.

The living portion of the coral is the thin fringe covering the surface of the structure. Here thousands of tiny animals feed and grow, secreting the limestone base that gradually builds the skeleton of the reef. These coral polyps may be golden yellow, or magenta, or vibrant blue or green. By day many polyps retract into their little limestone cups, but at night the coral really blooms. The polyps all emerge, spreading symmetrical tentacles like the petals of tiny flowers.

Coral grows in many different forms that frequently resemble various types of plants on land — a bunch of lettuce, a mushroom, a cauliflower, or a bush in bloom. And for a long time scientists assumed that coral was an aquatic plant. Then the surprising discovery was made that the coral polyp is a simple animal containing just two layers of cells. It feeds on plankton and other organic material in the sea. The tentacles that move and wave in the water seek out microscopic animals that float there. Each tentacle contains many coiled tubes as thin as threads. These nematocysts unwind with almost explosive speed and inject the prey with a poisonous fluid. Then the traumatized organism is passed along toward the mouth at the center of the "flower," which is always open, waiting for the next morsel.

Since the coral polyp is an animal, it must obtain its energy by the ingestion of organic molecules in which the sun's energy has been stored by plants using the process of photosynthesis. Capturing plankton from the sea is one way of obtaining this nourishment, but coral has another much more reliable food supply. It grows its private vegetable crop within the shelter of its own body. And the coral reef in its turn provides food and a home for millions of these tiny plants. Biologists conjecture that this marvelous symbiotic relationship began hundreds of millions of years ago when a living coral suffered an injury and the wound in its bony structure was invaded by single-celled algae of the type called zooxanthellae. Here they found a rich source of the nutrients they need — the waste produced by the metabolism of the coral polyps. They multiplied rapidly, packing the honey-combed coral structure with plant cells. The algae capture the energy of the sun, synthesize organic compounds, and thus provide an ideal source of food for the coral. The new relationship, beneficial to both the coral and the zooxanthellae, was so successful that it spread rapidly throughout the coral reefs. There are now hundreds of coral species that have their own special zooxanthellae living with them in this intimate relationship. But in order to raise a good crop of algae, the living portion of the coral reef must be near the ocean surface so the zooxanthellae receive a considerable amount of sunshine. A fast-growing reef is usually no deeper than 170 feet.

The coral polyp builds its bony skeleton by extracting calcium from the sea and combining it with carbon dioxide and water to make aragonite, a crystalline form of calcium carbonate that is harder and more resistant than other types of limestone. Most of this activity takes place during daylight hours because it depends upon the energy of sunlight. So the growth of the reef waxes and wanes in a regular rhythm day by day, and year by year, producing a visible pattern of rings in the stone — a beautifully detailed record of the sunlit hours on earth. Incredible as it may seem, it is possible today to read the fine print of this record back as far as 400 million years. It tells a fascinating story about the diurnal cycles and the relativity of time. Reefs that grew in the Devonian Period (nearly 400 million years ago) have 400 diurnal growth lines in each annual band. Somewhat younger corals (270 million years old) have 387 lines per year, and modern ones have about 360. These facts suggest a progressive decrease in the number of days a year and presumably an increase in the length of each day since the earliest reefs were formed.

The small, primitive coral polyp has a varied reproductive capacity that might well stir the envy of more advanced species. Genesis of new life most frequently takes place by budding. A miniature new polyp forms — an identical carbon copy of its parent — and remains attached to the parent as it grows. Eventually it also buds in its turn and so a colony of polyps develops. This cluster of organisms is an association for which there is no parallel in human life. More closely related than a family, the polyps remain attached to each other and have identical genetic endowment: a cluster of clones.

In fact, it is not clear whether "they" can in any real sense be called individuals. Do they feel and respond as separate organisms? Does each have a separate consciousness? (For surely consciousness is the central experience of life.) As our bodies grow and cells are added, consciousness encompasses the new growth; it is all gathered up into the same "me." But what happens in budding? Is a brand-new consciousness formed? Does rejuvenation take place at each budding so that the parent never really grows old?

In many ways the cluster of coral clones acts like a single complex individual — a superorganism. As division proceeds, the individual polyps line up side by side, building a bony structure characteristic of the particular species to which they belong. In some, the units are very tightly spaced. In others, each separate cup is surrounded by a broad foundation of limestone. There are hundreds of coral types, varying all the way from the densely packed whorls of brain coral to the open patterns of staghorn coral. Under given conditions of wind and wave and sunshine certain shapes are more favorable than others. In the rough waters of the open ocean, the coral heads that survive the longest are massive ones with compact rounded shapes that resist the pounding of the sea. In quiet, protected waters the delicate, branching forms offer more surface for intercepting the light and sieving plankton from the water. The skeleton of the reef also provides suitable living places for the zooxanthellae, arranging them so that they receive the proper amount of sunlight. Thus the individual coral polyps and their algae are working units in a larger and more complex structure that is advantageous to them all, just as primitive single cells joined with others of their kind eons ago to form the first eucaryotic organisms. In fact, the chloroplasts that are now an integral part of the zooxanthellae, enabling them to capture the energy of sunlight, may once have been free-living bacteria. Thus, step by step, nature has built higher levels of form, using individuals as the building blocks. As each organizational level matures, it tends toward an increasing interdependence of parts. The identity of the individual becomes more and more submerged until finally it can no longer behave as a separate entity. The chloroplast of the zooxanthella has completely given up its autonomy; it cannot survive alone. The coral polyp, on the other hand, still retains this option. It can exist and even reproduce itself as a separate individual.

In addition to asexual budding, the coral reproduces sexually, and each coral polyp has the marvelous ability to play either the male or the female role, switching sex apparently at whim. At certain seasons of the year sperm and eggs develop within a fold of internal skin. The sperm are released when they are ripe and drift until they are drawn into the mouth of a polyp containing mature eggs. The fertilized eggs soon produce tiny larvae covered with microscopic hairs. When these are released in their turn, they use the little hairs like arms to swim and guide themselves through the water. Amorphous as tiny clouds, they take on an almost infinite variety of forms. Drop-shaped one moment, they may be sausage or crescent-shaped the next. Millions of these larvae — or planulae, as they are called — emerge from a living reef at certain times throughout the year. In the Australian winter the breeding seasons coincide with the occurrence of the full moon. In summer the new moon provides the signal. No one understands how phases of the moon trigger this prodigal outpouring of new life when great drifts of reddish brown planulae stream from the reefs and are distributed by the winds and tides. They make long ugly stains across the otherwise clear water, looking unpleasantly like very thin sewage. A characteristic stench rises from them and fouls the sea breezes.

Embryonic life forms are seldom pleasing to the senses. Color, graceful form, and fragrance develop as an organism grows and usually reach their peak when the organism achieves the reproductive age itself, because beauty is an aphrodisiac. Once its purpose has been served, the life cycle repeats, raw new life appears, and ugliness returns.

～

Almost all of the planulae are eaten or die before they find a place where they can attach themselves and start to grow. They are bare and totally defenseless in a predatory world. Those that survive must encounter a very special set of circumstances before they can start a new colony. They cannot settle down on ordinary ocean bottom or they will be buried by shifting silt and sediment. In some mysterious way they seem to "know" that they must find a hard surface to which to attach themselves — a surface raised above the bottom of the sea. A rock, a submerged wreck, an empty portion of an old reef — these are the foundations they must use in order to begin their work of fashioning a new coral community. Any protrusions or irregularities in the ocean bottom provide an opportunity for the coral planula. It makes its choice and within a few hours has constructed its own hard, cuplike base, cementing itself to its new home. Inheritance has endowed the planula with an initial population of zooxanthellae, so it can immediately start to raise its own food supply. This tiny nucleus grows; it buds and steadily builds a more and more elaborate skeleton, adding identical units in a predetermined three-dimensional pattern very similar to the way a crystal grows.

It is interesting to observe how creativity in both the organic and inorganic world takes hold and blossoms on some small intruding form of matter. If all the ocean bottom were smooth and uniform, the coral planulae could not gain a foothold, just as a supercooled cloud does not start to crystallize into snow until a tiny seed or two is added — a particle of dust or sand. In the same way, cosmologists tell us, the universe itself began to take on shape and form after the Big Bang had scattered all matter helter-skelter into space. Random motion in that amorphous cloud caused local concentrations of matter. Once the concentrations existed, gravity could take hold. Stars condensed out of the whirling dust clouds, galaxies shaped up into vast rotating pinwheels of matter, planets were spun off, and molecules lined up in elaborate three-dimensional form. Just a small accumulation of matter — a knot in the grain of space-time — and from that everything else followed.

❧

As the new coral community divides and multiplies, it builds an underwater castle with many rooms and corridors providing shelter for its symbiotic zooxanthellae, and safe hiding places for the multitude of fish that are attracted to the growing reef. Soft-bodied gorgonians find the coral an ideal platform on which to erect their graceful fans where the sun shines through and the rough pounding of unprotected sea is broken and moderated. Crinoids, sponges, sea worms, and anemones take up residence in the sheltering, richly varied environment. They hide their long feathery tentacles in the little protected corners of the coral castle during the day. At night they open up, revealing their delicate beauty. Like a corps of ballerinas waving graceful arms, they skim microscopic organisms from the sea. In just a few decades a mature reef has grown where before there was nothing but an empty rock or a bare bone.

The bommie near Michaelmas Cay must have grown in just this way, flowering from some small protrusion on the ocean floor. As I floated facedown in the sunlit sea above the reef, I could observe at leisure its fascinating variety of shape and texture and motion. Near the center a deep crevasse went down to mysterious blue depths. Large fish moved into sight in this hole and then disappeared again. A little octopus slithered past, trailing its slate-blue tentacles. A sea urchin nestled beside a large brain coral, its dark quills stirring slightly as waves rippled across the reef. Close beside me — so near I might have touched it if I had tried — a pink and green parrot fish was busy nibbling at a branch of staghorn coral. Butting its small beaklike mouth against the branch, it nipped off little pieces of the living edge of the reef. The coral swallowed by the fish passes through its digestive system; the algae are absorbed as nourishment and the limestone is returned to the sea. These grains of sand filter slowly down through the water. They are washed and tumbled and polished by the waves. Gradually they pile up in great drifts on the ocean floor, or are thrown up onto coral cays and palm-fringed beaches. Much of the coral sand that is found around the world has passed through the digestive tracts of organisms like the parrot and butterfly fish, the bristlecone worm, and the starfish that regularly feed upon the coral reef.

To the casual observer the coral community seems to be a dazzling assembly of many beautiful individual things all competing for food and space and sunlight. But actually this galaxy of creatures has evolved intimate, closely interdependent relationships.

One of the most significant biologic discoveries in recent years is the importance of organized systems in nature. This insight came as a surprise to modern man, nurtured on the theory of survival of the fittest which emphasizes competition, the struggle for dominance, and survival at the expense of others. Cooperation and symbiotic relationships were seen as odd exceptions to the rule. Now we are beginning to understand that these integrating activities are essential to the way nature constructs ever more complex beings by putting together simpler things. The reef, including all the species that inhabit it, functions as one giant living unit. It is a superorganism on a larger and more sophisticated level than the coral colony. It feeds and grows and reproduces itself.

The coral skeleton provides the "bones," the supporting tissue for the giant organism. Its porous construction with many caves and tunnels acts as a circulatory system, allowing the life-blood of the sea to reach every "cell" of the body. The efficiency of the system is enhanced by the presence of sponges and many other organisms that burrow into the reef, creating an even more elaborate network of passageways and pumping water through the pores of their own bodies. During the day vast numbers of fish hover in dense schools over the reef, feeding on plankton and microscopic animals that float in the sea. The rain of fecal pellets dropped by the fish is consumed by the coral polyps. These contain phosphates and nitrates, which are excreted in turn by the polyps and passed along to the zooxanthellae, providing essential fertilizer for the coral's vegetable garden. Besides producing organic nutrients for the coral polyps, the algae also manufacture a limy deposit that helps to cement loose sand and sediment and repair minor damages to the skeleton.

Sea urchins perform an important maintenance function, too. They clean infected portions of the reef, feeding on the blue-green algae that grow on torn coral tissue and that might threaten the health of the reef. The coral skeleton is constantly trimmed and shaped up by butterfly fish and bristlecone worms. Excessive growth of any one part of the reef is also kept in check by a kind of chemical pruning. Inside each polyp are filaments that digest animal tissue and when these are protruded through holes in the body wall they can absorb the polyps of neighboring corals, preventing encroachment.

The reef has a communication network like a giant nervous system. Chemical signals carrying vital information about sex, food supply, injury, and impending danger are discharged into the water and circulated by the sea. As the subunits of the organism receive these messages, they are stimulated to respond in ways that coordinate the feeding, reproduction, growth, and regeneration of the larger whole.

Considering the remarkable integration of this great community of living things, it seems possible that our concept of organism should be broadened to include colonies of closely interdependent individuals. Although, strictly speaking, the coral reef is not a single living entity, it may be one in the process of formation. All in all, the reef functions so well as a cooperative venture that it is one of the most productive natural communities on earth. It yields more living matter per acre than carefully cultivated prime farmland.

❧

Swimming to the edge of the reef, I looked down and could see the enormous cliff of coral, dropping straight down below me into the dark abyss where light does not penetrate. I found a smooth coral head on which to stand and rested there for a moment, looking up at the cloudless sky. Far, far up there, I thought, light also gradually fades away and merges into the blackness of space. Two hundred miles above the earth the sky is as dark as those mysterious depths lying just twenty fathoms or so beneath my feet. This wonderful envelope of air and water surrounding the earth holds and distributes the sunlight, bathing everything that exists here in light from dawn to dusk. Only a narrow band of space holds this diurnal magic. Everything else is plunged in eternal night.

The water in which I stood was clear as air. I extended my arms and swam over the edge of the bommie, describing a long path out into the blue. Then I turned and caught my first breathtaking view of the whole wall of the reef, rising up from the dark depths.

At the lowest levels where only dim light penetrated, half-delineated shadowy shapes moved in and out of sight beneath clumps of brain and moon coral. As the wall emerged into more illuminated upper layers, it was clothed in increasingly delicate form and brilliant color. Lavender and gold fans waved gracefully in the ceaseless swells that stirred the water like a giant heartbeat. Schools of many different kinds of fish lived in this rich, active borderland between reef and sea. In colored clouds they hovered there, floating back and forth with the same rhythmic movement that bent the gorgonian fans. The crest of the coral wall was finely carved with leaves and branches and pinnacles, making a filigree whose exquisite detailing would have shamed the builders of the Taj Mahal.

If all the sea could be drained away, the enormous quantity, the complexity, and fantastic workmanship of the coral reefs would suddenly be revealed. These walls of lithosphere built by living things would stand forth, rising up dramatically from the ocean bottom — not in one solid, continuous surface but in many free-standing pieces, like skyscrapers and apartment complexes along a maze of city streets.

249

The sheer mass of coral reefs staggers the imagination; the Great Barrier Reef alone covers 80,000 square miles.

The presence of these great coral reefs poses an interesting question about which scientists have argued ever since Darwin visited the Great Barrier Reef in 1835. Because coral thrives only in clear seawater between 75° and 95° F and not more than 170 feet deep, the land off the east coast of Australia must have been nearer the surface twenty thousand years ago when the lowest levels were formed. Now the bases of the reefs are many hundreds of feet below water level. Did the land subside or did the ocean rise? Darwin believed that the land slowly sank and as it did, the coral built ever higher and higher ramparts to intercept the sun. Other scientists, however, have pointed out that ocean levels were lower when the great ice ages held more of the earth's water locked up in glaciers. As the glaciers melted and the waters slowly rose, the coral reefs grew upward, keeping pace with the rising sea levels. This latter hypothesis is favored today as the explanation of the Barrier Reef, but Darwin's theory of subsidence is needed to account for the coral atolls where reefs form rings around volcanic cones that have sunk far beneath the surface of the sea.

The size and location of reefs provide much valuable information about earth history, going back as far as 450 million years ago when the oldest forms of coral flourished in early Paleozoic oceans. Although these ancient coral species are slightly different from the ones that exist today, biologists are reasonably certain that the range of ideal conditions for the ancient corals was also very restrictive and similar to the optimum conditions at the present time.

Fossil corals have been found in the most unlikely places. They are imbedded in the snow-covered rocks of the Himalayas, 20,000 feet above sea level. They have been found on the windswept slopes of the Alps, where temperatures often drop to 50° F below zero. They have been found in the forbidding Tassili Mountains in the heart of the Sahara Desert, where only a few inches of rain fall in a year and the nearest ocean is 500 miles away.

The distribution of fossil corals reveals the location, the size and shape, the temperature and salinity, of ancient tropical seas. Similarities and differences in coral species as they developed throughout time document the changing geometry of land and ocean basins. For example, about three million years ago the isthmus of Central America was born by volcanic action. The closing of direct sea connections between the Atlantic and Pacific can be dated by evidence left by coral polyps in the Bahamas and Baja California. Species that had been similar for millions of years became widely divergent as the land barrier formed and separated them.

So the coral reef has been patiently at work eon after eon altering the profile of our planet. Seemingly fragile but strong with the resilience of life, the reefs have protected the shoreline of continents from the violence of the sea. They have formed little islands like Michaelmas Cay. They have circled volcanic cones in tropic waters, extending their area, sheltering them from erosion by the long rollers, the tsunamis, and the typhoons that sweep across these seas. From age to age, they have laboriously constructed acres of new earth crust — a living portion of the lithosphere. Even in death the coral polyp contributes its bit to the evolving shape of the planet. Fragments of its skeleton, broken up into grains of sand, lie in great drifts on the ocean bottom or are washed up onto the edge of continents, making the cool, glistening coral beaches around the world. No other organism — even including man — has made such an important contribution to the crust of the earth.

~~

As I swam in deep blue water back toward the diving boat, I was surrounded by a vast school of pale, transparent little fish. Racing toward the reef, they seemed oblivious of my presence. And so thick were they in the water that I expected to feel them brush against my skin. But not one of them touched me and in a minute they had disappeared within the protective chambers of the reef.

Were these fish fleeing from danger, I wondered? And, looking around, I saw a large gray reef shark swimming just a few yards behind me. I turned and faced the shark. (Not that I was brave — it seemed worse to have it following me!) The shark stopped swimming also and looked at me, shaking its head in a jerky, nervous movement. The eyes twitched as the creature took in my odd appearance. Treading water there in my orange wetsuit, I must have looked like a giant clownfish caught out away from its anemone.

Time came to a standstill as the shark and I stared at each other. Many thoughts tumbled through my head, scraps of information and cautionary instructions I had heard over the years: "If you sight a shark, swim strongly. Don't give the impression of a weak or wounded fish. . . . On the other hand don't swim too fast, the frantic rhythm of your movement may act as a trigger. . . . Sharks' brains are triggered by some sound or shape — even the silhouette you present in the water might trigger an attack. . . . (What kind of shape should I present?) Gray reef sharks are not very fast; they have trouble catching fish, so they are always hungry. . . . Don't make any sudden offensive gesture; don't strike at a reef shark or it will surely attack you. . . ." All these disorganized thoughts flickered through my mind but none of them really supplanted the vivid memory of the photograph showing a whole sea turtle in a shark's stomach.

Although the confrontation could not have lasted more than a fraction of a minute, it seemed infinitely long. Then suddenly the shark flipped its tail strongly and swam off, its shining, streamlined body gliding gracefully through the water.

The shark disappeared into the dim recesses of its underwater world and I returned to the world of man. But vivid memories of the life beneath the sea remained with me. As I trod the concrete streets of the cities and rode the elevators up the towers of glass and steel, I thought about the many-chambered and multistoried reef. We human beings, too, have built intricate and marvelous dwelling places, hard and enduring like the coral castles. But our cities are not places where life blooms. There is no living community here, no rich diversity of other living forms. Oh, yes, a few dogs and cats roam our streets; rats hide in dark, noisome corners; pigeons crowd the cornices and statues of our public squares. But these are just the remnants, the residue left over in a general retreat.

Four hundred and fifty million years of evolutionary progress lie between the invention of the coral reefs and the cities of man. Much has been gained in those millennia. The mind has evolved, and language has been discovered. Music and literature have burgeoned and many secrets of nature have been unlocked. While all this has been happening, some of the old instincts have been preserved. Self-defense and the predatory drives are still strong within us, as witnessed by the locks and bolts on all our houses and the screams that issue from dark alleys in the night. They are there undiluted — brutal and elemental — as they are in the shark.

But one marvelous adaption has slowly dropped away: the ability to take part in symbiotic relationships, to cooperate in building larger, more advanced organizations of living matter. Almost everywhere in nature, the presence of one life form provides a toehold, a refuge, a rich pasture for other living things. Where there is life, other life flowers. And each form provides for the other in curious and wonderful ways. Not so man. I wonder where in the long evolutionary journey this gift began to slip away, and whether it will prove to be a fatal loss in the end.

14

The Whole Earth

The greatest beauty is organic wholeness,
the wholeness of life and things.

— ROBINSON JEFFERS

I was born too soon to ever see with my own eyes the whole earth silhouetted against the dark abyss of space, or coming in closer to watch the continents emerge from their veil of clouds, revealing the sculptured profile of their coasts against the shining darkness of the deep blue sea. But today I have seen a small piece of the earth — a tiny microcosm — from the perspective of space, as I soared on the trade-winds above the island of Oahu.

On this jewel-like Hawaiian morning I was towed aloft in a sailplane from a field on the northern shore of the island where the Waianae Range meets the sea. The breeze came clean and steady out of the northeast after traveling uninterrupted over thousands of miles of open water. Encountering the ridge of mountains, it swings upward in a wave of rising current; so when I released from tow I was borne aloft as a gull rides the ascending winds. Smooth as silk the air rose, carrying me up over glistening coral beaches, over velvet green mountain slopes, over cobalt sea — higher and higher until I seemed suspended in a sunlit bubble of space and a great quietness — the voice of the wind muted to a whisper as it flowed gently over the wings.

The mountain slopes were close below me and the dense vegetation, which had looked like a carpet of moss from a distance, was differentiated into many forms and shades of green — dark diffuse groves of ironwood, spreading umbrella canopies of monkeypod trees, giant *koa*, and banyan festooned with curtains of *koali* vine. Above the soft understory of fern trees, the dancing leaves of the *'ōlapa* trees and tasseled tops of lobeliads trembled in the breeze. Deep in the hollow of each ravine rows of pale *kukui* trees made long wandering ribbons like glacial streams flowing down the dark lush valleys to the sea.

As I followed the ridge westward, I could catch glimpses between the jagged peaks of the land beyond the hills. And then where the mountains descended suddenly to the shore the whole leeward coast came into view: valleys bathed in sunshine, mile upon mile of pineapple plantations with young plants set in neat rows against the mahogany-red Hawaiian soil, and vast seas of silvery green sugarcane fields ripe for harvest.

The rising wave of air crested at about 2,500 feet, but soft cumulus clouds were floating near the ridge. The air, rising up to meet each one of these, made small eddies of ascending current; and I sailed from cumulus to cumulus receiving a lift from each one. In the language of glider pilots, I was soaring down a "cloud street."

Soon I could see the far coastlines of the island: the south shore with the three deeply indented lobes of Pearl Harbor; the east shore, soft and misty, veiled in clouds hanging low upon the dark hulking shapes of the Koolau Range. The whole island was there below me — complete and self-contained — a tiny model of the earth in time as well as in space. This little island can be seen and understood in a way that is not yet possible for the entire planet. We know its past — for it is very young — and we can guess its future.

Oahu is one of a long chain of volcanic islands trending north–northwest across the Pacific. At the southern end of the chain the island of Hawaii is actively volcanic today. On Maui, the next island to the northwest, the volcanoes are extinct but only slightly eroded. Then come Molokai and Oahu with volcanoes long extinct and deeply eroded. Kauai, Nihoa, Necker, Lisianski, Midway, and Kure Atoll lie in a sequence to the northwest. Beyond Kure stretches a chain of drowned islands, guyots, and seamounts that rise steeply from a broad submarine ridge. The ages of the islands have been determined and they fall in the same order. Hawaii is the youngest (recent to half a million years), Oahu is four million years old, Kauai is about five million years, Midway is eighteen million years old, and Kure Atoll is twenty million. This regular succession suggests that the Hawaiian chain formed as the plate carrying the Pacific Ocean passed in a northwesterly direction over a plume of concentrated heat in the earth's mantle. Twenty million years ago the plume lay beneath Kure Atoll and now Hawaii is located over this hot spot.

These new bits of earth crust, emerging suddenly from the sea, were far removed from the other landmasses. For many months the black rocks were stark and barren as the moon, devoid of all life. Then as blazing tropical sunshine alternated with drenching rains, the harsh surfaces slowly began to soften. Winds brought a varied assortment of life forms. Spiders sailed in on strands of silken webs; these tiny voyagers are frequently seen miles above the earth's surface in such great numbers that at times the air takes on an iridescent sheen and they have been among the first forms of life to take up residence on islands newly born out of the sea. Seeds and spores light enough to float on the breeze were carried thousands of miles from more ancient lands and deposited at random across the bare mountain flanks. One out of a hundred million, perhaps, found a toehold on the dark, forbidding rocks. But these grew and began to work the magic of vegetation upon the land. Lichens were probably the first successful flora. Like the coral colony, lichens are not single individual plants; they are a symbiotic combination of algae and fungi. The algae capture the sun's energy by photosynthesis and store it in organic molecules. The fungi absorb moisture and mineral salts from the soil and rocks on which they grow, passing these on as waste products that nourish the algae. Lichens helped to speed the decomposition of the hard rock surfaces, preparing a soft bed of soil abundantly supplied with minerals that had been carried in the molten magma from the inside of the earth. Now other forms of life could take hold: ferns and mosses and low-growing shrubs that flourish even in rocky crevices. Plants with buoyant seeds — coconuts and beanlike pods — drifted on the ocean currents and were washed up on the shores. Remarkably resistant to the vicissitudes of ocean travel, they could survive prolonged immersion in salt water. When they came to rest on warm beaches, a few of them sprouted and grew. Their descendants are now among the most abundant trees on the island, highlighting the dark tropical forests with bright flowers or casting graceful fringed shadows along the edge of the sea. Birds swept off course by the storms found a safe haven on these shores and more seeds hitched a ride on the birds' wings. Hibiscus plants, cotton, and *koa* trees were among the earliest colonists. Flowers found the new islands an ideal habitat; nurtured in the tropical warmth and brilliant sunshine, they grew to enormous size. A paradise of lush beauty took shape where only a little while ago dark lava had erupted from a boiling sea.

Before the advent of man the Hawaiian Islands were so isolated that the plants and animals developed in unique ways, producing species that are found nowhere else. Hawaii has the only carnivorous caterpillar in the world. Wingless and blind insects evolved here. Some birds, like the rails, found such an abundant ground-based food supply that they had no need to travel far, and they, too, lost the power of flight. Land snails became tree snails. Ferns, shrubs, and flowers, finding little competition from larger vegetation, grew tall and produced unusual varieties of trees like lobeliads and tree violets. On the other hand, several familiar types of life were conspicuously missing. No truly fresh-water fish evolved in the Hawaiian Islands, no native amphibians, reptiles, or mammals except one species of bat. Large insects were not successful in colonizing the islands, but large flowers were abundant and special types of birds evolved to take over the role of pollination. From one original progenitor there are now twenty-three species of honeycreepers, dainty birds with long, pointed bills to sip the nectar from the angel's trumpet, the *'ohé naupaka*, the cup of gold. A unique and closely interrelated colony of living things grew and flourished, as rich and colorful, as finely integrated as a coral reef. And while all this was taking place on land, coral polyps were busy constructing their elaborate communities on the outcrops lying just beneath the surface of the sea, forming a fragmentary fringing reef along the shores.

Geologically, we would expect the future of Oahu to follow the usual development of a coral atoll. Although isolation has limited the variety of coral species (there are only 42 in Hawaii as compared to 300 species in some parts of the world), if nature were allowed to take its course the isolated reefs would grow and finally form a nearly

continuous structure. The land, becoming more compacted and eroded, would subside. The coral reef would grow higher as its base sank deeper, and soon a barrier reef would enclose a shallow lagoon between itself and the island. Finally the island would sink entirely beneath the water, leaving a ring-shaped coral atoll cupping a little segment of the sea.

Models for this evolution exist in a number of places in the Pacific. The Society Islands, for example, have passed quite rapidly through the first portion of this scenario. Bora Bora, which is almost an atoll, having just a small volcano core surrounded by a barrier reef, is only four million years old, the same age as Oahu. But many fully developed atolls, such as Bikini, Eniwetok, Kwajalein, Wotje, and Ailuk, are very old. The age, depth, and structure of Eniwetok and Bikini were subjected to careful scientific study in preparation for the nuclear tests conducted there in the 1950s. Eniwetok's reef was found to extend down three quarters of a mile below sea level where it rests on the top of a volcanic cone whose crest is two miles above the sea floor. The age of the rocks reveals the atoll's development throughout time. Fifty million years ago the volcano sank below sea level and the barrier coral reef become a circular ring in the sea. For the next twenty-five million years the volcanic cone continued to subside at the rate of about two inches per thousand years. Then the rate of subsidence slowed down and for the last ten million years it has averaged about half an inch each millennium. All this time the coral reef kept pace, diligently adding to its stature in order to maintain its favored place in the sunlit shallows of the sea.

If we make the assumption, which seems to fit the facts very neatly, that the Hawaiian archipelago has moved northwest on the Pacific plate at the rate of about three inches a year, then we can understand the history of these islands as they have aged, subsided, grown a necklace of coral, and finally passed into latitudes unfavorable for reef formation. As the islands moved north into colder waters, coral growth slowed down but over a certain range of latitudes it was still rapid enough to compensate for small, gradual changes in depth. Kure Atoll lies at about 28 degrees north. Beyond that point the islands were drowned because the reef was not able to keep pace with subsidence and rising sea levels. Following this sequence, Oahu should become a coral atoll in about six to seven million years while it is still within waters warm enough to maintain vigorous reef formation. However, there is one factor that could alter this prognosis — the impact of man.

On the eastern side of the island at the foot of the Koolau Range, Kaneohe Bay receives the drainage from ten rivers that cascade down the precipitous mountainsides. When heavy rains fall, the swollen rivers carry a thick load of red volcanic soil into the sea. The island seems to be bleeding into Kaneohe Bay. Land has been cleared along this coast for housing and highway construction, making it more susceptible to very rapid erosion, and for many years sewage outfalls fouled the waters of this bay, which was once the site of spectacular reefs. Known as the Coral Gardens, this formation was one of the special treasures of the Hawaiian Islands.

Coral reefs have a natural mechanism for cleaning the sediment that falls on their surfaces. A thin coating of mucus secreted by the coral polyps traps the particles, which are then moved down and off the reef by millions of waving cilia. But this process cannot deal effectively with a thick layer of sediment. The cilia are unable to move the heavier load. Then the zooxanthellae, deprived of light, die, and the reef loses its principal source of nourishment.

Sewage damages the dynamic integrity of the living reef in another way. Nutrients contained in the effluents encourage the growth of algae that multiply prodigiously and can no longer be kept in check by the natural defenses of the reef community. In Kaneohe Bay green bubble algae took over like a spreading cancer engulfing entire coral heads. Continuous sheets of this thick warty growth enveloped broad areas of what was once a magnificent piece of living lithosphere. The smothered coral polyps died and even the limestone began to dissolve beneath the algae. Recognition of these problems led to corrective measures that are beginning to restore the health of the reefs. Sewage has been diverted to the open ocean and grading ordinances have reduced erosion.

In other ways, however, human beings are continuing to make serious inroads on the Hawaiian reefs. Aggressive fishing methods are responsible for great areas of dead reef. Clorox is sprayed into the coral caves and recesses where aquatic organisms make their homes, and they are driven out into nets in vast numbers. The reef that has been sprayed never recovers.

Pieces of coral are removed from the reefs for building materials and tourist souvenirs. As a result of these human interventions, the coral communities may never be vigorous enough to grow into a solid barrier and create an atoll before the island is carried northward into waters so cold that the growth cannot keep pace with subsidence. The land itself, cleared of vegetation in many places and exposed to erosion, will be reduced to sea level much more rapidly than would occur in an undisturbed environment.

The changes wrought by human activities proceed subtly at first by small, hardly noticeable increments from one year to the next. But, long continued over centuries and millennia, they can disrupt the organization of the island and even alter its future. The soil, the air, the water, and the envelope of living organisms are all part of a single evolving whole. Excessive growth of any one part is usually controlled by other members of the community, but nothing is powerful enough to control man except man himself.

❧

Viewed from the perspective of my sailplane, the island of Oahu is still various and beautiful, lifting strong, rocky shoulders out of an azure sea. Lighted by shafts of sunlight, the fields are vibrant with color of an almost unbearable intensity like a green flash at sunset.

The glider port is below me now between the mountains and the shore. To lose altitude I leave the ridge behind and soar out over the blue water, across the white-crested waves that mark the margins of the sea like bars of music. The view from my plane is almost all-encompassing. With only a slight dip of one wing I can see the whole circle of the horizon. I can look down on the sapphire and gold ocean and up at the frothy clouds and azure sky arched above my head. Skimming here in the luminous blue envelope of air and water that wraps the earth, I experience a freedom that no previous generations of human beings have ever known. I can explore the depth, breadth, and heights of space. On the wings of sailplanes and jets, in rockets and satellites, mankind has broken loose from the shackles that bound him to the planet's surface for millions of years. Above the cloud tops, beyond the aura of light, men have looked back on the earth, seen it whole and beautiful as I have seen this island today.

But mankind is still closely confined to a tiny segment of the time dimension — three-score years, perhaps, or four. Beyond that boundary he can pass only in imagination. "Like the wistaria on the garden wall," Loren Eiseley said, "he is rooted in his particular century. Out of it — forward or backward — he cannot run. As he stands on his circumscribed pinpoint of time, his sight for the past is growing longer, and even the shadowy outlines of the galactic future are growing clearer, though his own fate he cannot yet see. Along the dimension of time, man, like the rooted vine in space, may never pass in person. Considering the innumerable devices by which the mindless root has evaded the limitations of its own stability, however, it may well be that man himself is slowly achieving powers over a new dimension — a dimension capable of presenting him with a wisdom he has barely begun to discern."

Traditionally we have equated time with the repetitive cycles that mark the days, the seasons, the years of our lives. As we focus on these familiar rounds, watching the sun and moon return over and over again to the same place in the heavens, it is natural to think, *Plus ça change, plus c'est la même chose*. We take comfort in the endless repetition of summer and winter, dawn and dusk — like a child on a merry-go-round who is reassured each time his pony comes back to the starting place and he sees his mother waving by the entrance gate. He does not know and would not believe it if he were told that during the half hour he has been whirling around and around, his mother, his pony, and indeed the whole earth have changed and moved great distances. The merry-go-round has traveled about 500 miles as the earth spun on its axis. The earth as a whole has rotated 34,000 miles on its orbit around the sun, and 247,500 miles in its trip around the Milky Way. And the galaxy has moved several hundred thousand miles in the direction of the constellation Virgo. The earth is not the same as it was half an hour ago. The midocean ridges have grown a little wider. The Rocky Mountains and the Andes have pushed a little higher into the sky. New coral reefs have been spawned. Perhaps a species has become extinct and others have undergone mutations, giving birth to some new life potential. Although there are cycles within cycles within cycles, none of them brings us back to the same place where we began. The repetitions provide only a convenient measure of time, like a metronome marking out the rhythm of our lives. Without this measure we would be truly awash in the turbulent ocean of cosmic change. On the other hand, without real change, the monotonous tick-tock of the days and years would go on forever measuring — nothing. The history of the earth has shown us that great changes have taken place. The planet has a past and a future. Someday we may be able to understand its biography just as we know the life story of Oahu.

In imagination, at least, we can achieve some perspective on the earth's development by speeding up the history of the planet. Let a billion years, for example, pass by in a single day. At the beginning of this time-lapse show we would see the infant planet emerge from the whirling dust of the solar nebula — a diffuse ball of gases and dust, enveloped in a kind of afterbirth, the primordial atmosphere of very light gases. Gradually this envelope drifts away into space and at the same time the material of the embryonic planet is drawn strongly toward the center under the force of gravity. So much heat is generated by this "falling of matter" that the planet becomes very hot; the elements and mineral compounds melt. The densest ones flow to the center and the less dense ones rise to the top, like foam on a boiling kettle. The surface of the gradually compacting ball is constantly peppered with more matter from space as the remaining particles from the solar nebula are attracted by the gravitational field. But still the planet cools; the solid crust begins to take shape, and even as it forms it is disrupted repeatedly by violent eruptions and incandescent flows of glowing lava. Gases are emitted, taking the place of the primordial atmosphere that is floating away. Clouds of water droplets condense in the atmosphere and rain falls; oceans begin to collect on the planet's surface; and soon living molecules are present in its shadowy depths. Exactly when they appear we cannot say. But it is a phenomenon of great importance because the creation of these earliest forms of life involves a higher degree of organized activity than anything that has previously occurred on our planet. The simple self-replicating molecule contains hundreds of thousands of atoms engaged in the multitudinous coordinated activities that make up a single organism. And all of this takes place in the first day after the birth of the planet.

By the early dawn of the second day the first cellular organisms have appeared and photosynthetic algae are forming vast colonies in the sea and altering further the composition of the fluid envelope — the membrane — that wraps the planet. Throughout this whole process the body of the earth is never still. It is constantly readjusting and reorganizing the materials of which it is composed. The very skin of the planet is being continuously reworked and made anew, like a snake's skin grown and cast off many times to make room for a new one; the ocean floor is melted down and reformed about five times a "day." The dry land is being altered, too, as mountains are lifted up, then swept away by wind and rain. While all these dramatic movements are occurring, less conspicuous changes are quietly proceeding beneath the surface. Deposits of copper and silver and gold are secreted in the deep roots of volcanic mountains and beautiful crystals are taking shape inside the rocks — diamonds and star sapphires and blood-red rubies. The second and the third day have come and gone.

The composition of the aura of gases surrounding the planet continues to evolve, enriched by the by-products of living things that are multiplying in the seas and by the flow of gases from volcanic eruptions, fumaroles, and warm sulfur springs. The chemistry of the oceans is changing, too, as hydrothermal vents pour forth their hot, mineral-laden waters. Four days and ten hours after the beginning, life begins to invade the land, profoundly altering the surfaces, and soon the envelope of living matter becomes complete around the whole earth, making an inner membrane for the planet, creating a structure of much greater complexity and infinitely greater potential. Floods of living things appear and disappear, but life itself continues to burgeon, moving to ever higher, more organized beings. Finally, in the last minute man appears, and in less time than it takes to read this sentence the new life form spreads across the surfaces of the earth, altering everything in its path.

With geologic time compressed into five days, the whole biography of the earth can be taken in at one glance, so to speak, and its dynamic quality stands forth, clearly revealed. Not just a haphazard collection of elements drawn together by chance from the solar cloud, the earth is an integrated whole with a form and a personality that have evolved since its birth and are still unfolding. Perhaps it is a superorganism in the process of formation, like the coral reef or the coral atoll. Or we might compare it with the assembly of protein molecules that became linked to each other in intimate, interdependent ways to make the first living cell. At so many levels of complexity nature provides us with models of organisms in the process of becoming.

The time-lapse history of the planet is evolutionary in the deepest sense and yet it is also reminiscent of the biblical story of creation, which was set down long before there was any comprehension of the immensity of geologic time. No one, even then, imagined that the formation of this magnificent world in which we live was an instantaneous act. It took place in time — be it six days or five billion years. Time is the dimension of creation.

How does man fit into this evolving system? That is the greatest mystery of all. In one respect his appearance represents a quantum jump toward higher levels of organization because the human mind is the most elaborate and delicately balanced integration of matter known in the universe. Lewis Thomas has suggested that "man may be engaged in the formation of something like a mind for the planet." This is an inspiring prospect. But so far, man has shown little inclination for collaboration with the rest of nature. He has profoundly disturbed the finely constructed systems of soil and sea, of sunlight and living things, that have developed over the eons of geologic times.

Such doubts probably underestimate the remarkable possibilities buried deep within matter. As John Muir observed, "One is constantly reminded of the infinite lavishness and fertility of Nature . . . no particle of her material is wasted or worn out. It is eternally flowing from use to use, beauty to yet higher beauty."

Even the humblest and least loved of all earth substances, the clay of the fields, has a miraculous and hidden use — the capacity to act as a mold, a template for the precursors of life. Perhaps we, too, contain within us powers yet untapped, powers that may enable us to fulfill some role above and beyond ourselves, to become an integral part of the whole earth, and to join with the rest of nature in creating a world new and wonderful beyond our imagining.

~~

In the photographs of the earth from space the planet looks like a little thing that I might hold in the hollow of my hand. I can imagine it would feel warm to the touch, vibrant and sensitive. Born of stardust, this handful of matter has evolved throughout the eons of geologic time. Like a butterfly taking shape within its chrysalis, the parts have rearranged themselves, taking on new forms. Diversity has increased, and simplicity has given way to elaborately integrated complexity. Beneath the mobile membrane of cloud and air are a storehouse of splendors and a wealth of delicate detail. There are rainbows caught in waterfalls, and frost flowers etched on windowpanes, and drops of dew scattered like jewels on meadow grass, and honeycreepers singing in the jacaranda tree.

Time flows on . . . the planet continues to spin on its path through the unknown reaches of space. We cannot guess its destination or its destiny. This beautiful blue bubble of matter holds many wonders still unrealized and a mysterious future waiting to unfold.

Table of Geologic Time

Era	Period (or Epoch)	Years before the present
CENOZOIC	Pleistocene Epoch	recent–2 million
	Tertiary Period	2–65 million
MESOZOIC	Cretaceous Period	65–136 million
	Jurassic Period	136–190 million
	Triassic Period	190–225 million
PALEOZOIC	Permian Period	225–280 million
	Carboniferous Period	280–345 million
	Devonian Period	345–395 million
	Silurian Period	395–430 million
	Ordovician Period	430–500 million
	Cambrian Period	500–570 million
PRECAMBRIAN		570 million–4.6 billion

Notes

Chapter 1

10   *Casanova quotation*: from Casanova de Seingalt, *Memoires*, as cited by White, "The Chinese Landscape," p. 97.

10   *Darwin quotation*: from Darwin, *The Voyage of the Beagle*, Chapter XV, p. 326.

11–12 *Matthiessen quotation*: from Matthiessen, *The Snow Leopard*, pp. 179–180.

12   *American Mount Everest expedition*: from Dyhrenfurth, "Six to the Summit," pp. 460–477.

12–13 *Traverse of Mount Everest:* as described by Hornbein and Unsoeld, "The First Traverse," pp. 509–514.

13–14 *Yellow band*: as described by Hornbein and Unsoeld, "The First Traverse," p. 510. Fossils and dating of the band as reported by Gansser, *Geology of the Himalayas*, pp. 164–165.

14   *Darwin's fossil discoveries*: as reported in Darwin, *The Voyage of the Beagle*, chapter XV, pp. 323–324 and 335–336.

15   *Fossils forming from seeds*: theory reported by Rudwick, *The Meaning of Fossils*, p. 84.

15–16 *Leonardo da Vinci's theories*: as reported by Rudwick, *The Meaning of Fossils*, pp. 39–40.

16   *Hooke's theories on fossils*: from Hooke, *Micrographia*. Also reported by Porter, *The Making of Geology*, p. 73.

16   *Hooke's view of earth history*: as described by Porter, *The Making of Geology*, pp. 82–83.

16–17 *Cuvier's theory*: as reported by Rudwick, *The Meaning of Fossils*, pp. 101–145.

17   *Lyell's uniformitarianism*: from Lyell, *Principles of Geology*. Also as described by Wilson, *Charles Lyell, The Years to 1841*.

18   *Rates of raising and leveling of land forms*: as cited by Strahler, *Principles of Physical Geology*, p. 297.
      3,500 Peruvians Perish in Seven Minutes."

19   *Dyhrenfurth quotation*: from Dyhrenfurth, "Six to the Summit," p. 472.

21–22 *Account of avalanche in Peru*: as described by McDowell, "Avalanche!

22   *Depth of sedimentation on ocean floor and rates of erosion*: as cited by Dott and Batten, *Evolution of the Earth*, p. 136.

Chapter 2

page

27   *Anthias a good augury*: as reported by de Latil, *Man and the Underwater World*, pp. 46–47.

27   *Aristotle quotation*: from Aristotle, *History of Animals*, IX, 25.

27   *Conditions encountered in sponge diving*: as reported by de Latil, *Man and the Underwater World*, pp. 46–47, citing Aristotle, *History of Animals*.

27–28   *Oppian quotation*: from Oppian, *Halieutics*, III, as cited by de Latil, *Man and the Underwater World*, p. 47.

28   *Xerxes' attack on Greece*: as described by Durant, *The Life of Greece*, pp. 237–239, citing Herodotus and Plutarch.

28–29   *Story of Scyllias and Cyana*: as described by Herodotus, *The History*, VIII, 8, and Pausanias, *Description of Greece*, X, 19. The two accounts somewhat conflict in regard to the timing and location of these events. I have followed in general the account of Pausanias.

29   *Xerxes punishes the sea*: as described by Herodotus, *The History*, VII, 35.

29   *Aristotle's description of cauldron*: as cited by de Latil, *Man and the Underwater World*, p. 43.

30   *Leonardo da Vinci's diving inventions*: from Leonardo da Vinci, *The Notebooks of Leonardo da Vinci*, pp. 161–164.

30   *Discoveries of Paul Bert*: as reported by Earle, *Exploring the Deep Frontier*, p. 104.

31   *Deep-sea drilling*: as described in *Rand McNally Atlas of the Oceans*, pp. 49, 165.

36–37   *Manganese nodules*: as described by Friedman and Sanders, *Principles of Sedimentology*, p. 140. Also *Rand McNally Atlas of the Oceans*, pp. 102–103; and Hopson, "Miners Are Reaching For Metal Riches on the Ocean's Floor," pp. 51–52.

38–39   *Continental shelves*: as described by Marden, "Man's New Frontier," pp. 502–507. Also *Rand McNally Atlas of the Oceans*, p. 22; and Wolff, "Land Beneath the Sea," pp. 34–41.

40   *Depth of light penetration*: as cited in *Rand McNally Atlas of the Oceans*, p. 26.

41   *Project FAMOUS*: as described by Heirtzler, "Project FAMOUS," pp. 586–603. Also *Rand McNally Atlas of the Oceans*, pp. 88–89.

41   *Van Andel quotation*: from Van Andel, *Tales of An Old Ocean*, p. 156.

41–42   *Dive 526 by Alvin*: as reported by Ballard, "Dive Into the Great Rift," p. 610.

42   *Minerals from mantle*: as reported by Ballard, "Window on Earth's Interior," pp. 247–248.

42–43   *Galapagos rift*: as described by Corliss and Ballard, "Oases of Life in the Cold Abyss," pp. 441–453.

44   *Vents off Baja California*: as reported by Waldrop, "Ocean's Hot Springs Stir Scientific Excitement," pp. 30–33. Also Sullivan, "Deep-Sea Life is Found Flourishing," p. 33; and Cooke, "Ocean of Discoveries Excites Scientists on Seabed," p. 26.

45   *Age of ocean water*: as stated by Schopf, *Paleoceanography*, p. 7.

Chapter 3

page

49   *Cox quotation*: from Cox et al., "Reversals of the Earth's Magnetic Field," p. 53.

49   *Lucretius quotation*: from Lucretius, *De Rerum Natura*, Book VI, lines 906 ff.

50   *Eratosthenes' experiment*: as described in *Encyclopaedia Britannica* (1956), vol. 8, p. 680.

51   *Pytheas's voyage*: as reported by Freuchen, *Peter Freuchen's Book of Arctic Exploration*, pp. 59–72. Also *Encyclopaedia Britannica* (1956), vol. 18, p. 804.

51   *History of compass*: as reported by Butler et al., "History of the Mariner's Compass," pp. 168–169. Also Bozorth, "Magnetism," p. 637.

52   *Neckam quotation*: as cited by Butler et al., "History of the Mariner's Compass," p. 169.

52   *Arabian writer quoted*: Bailak Kibdjaki, *Merchant's Treasure*, 1282, as cited by Butler et al., "History of the Mariner's Compass," p. 169.

52   *Medieval navigators awed by magnetism*: as described by Bozorth, "Magnetism," p. 637, citing as source Petrus Perigrinus de Maricourt, 1269.

52   *Mariners' idea of magnetic pole*: as cited by Freuchen, *Peter Freuchen's Book of Arctic Exploration*, p. 82.

52   *The pilot's blessing*: described by Morison, ed., *Journals and Other Documents on the Life and Voyages of Christopher Columbus*, p. 103.

53   *Columbus's experience with compass deviation*: described by Morison, ed., *Journals and Other Documents on the Life and Voyages of Christopher Columbus*, p. 203. Also de Madariaga, *Christopher Columbus*, p. 202.

54   *Columbus correcting for deviation*: theory reported by Parasnis, *Magnetism from Lodestone to Polar Wandering*, p. 92.

54   *Accuracy of log*: as reported by Morison, ed., *Journals and Other Documents of Christopher Columbus*, pp. 203–204. Also de Madariaga, *Christopher Columbus*, pp. 191 and 205.

54–55 *Magnetic deviation in Atlantic*: as reported by Parasnis, *Magnetism from Lodestone to Polar Wandering*, p. 102.

55–56 *Gilbert's work*: as reported by Parasnis, *Magnetism from Lodestone to Polar Wandering*, pp. 11 and 103–107.

56   *Electromagnetic discoveries*: as described by Elsasser, "The Earth as a Dynamo," p. 44.

57   *Geomagnetic reversals discovered*: as reported by Cox et al., "Reversals of the Earth's Magnetic Field," p. 44.

59–60 *Vine at Lamont-Doherty*: episode related by Frank Richter, University of Chicago, personal communication, February, 1980.

60   *Pattern of reversals in geomagnetic field*: as reported by Heirtzler et al., "Marine Magnetic Anomalies," p. 270. Also Wyllie, *The Way the Earth Works*, pp. 130–132.

60–61 *Recent changes in geomagnetic field*: as interpreted by Wyllie, *The Way the Earth Works*, p. 132.

61   *Gothenburg Excursion and climate changes*: as described by Fairbridge, "Global Climate Change During the 13,500 B.P. Gothenburg Excursion," pp. 430–431.

61   *Description of last ice age*: as reported by Kukla, "Around the Ice Age World," pp. 56–61.

page
62   *Magnetic reversals and climate*: as presented by Fairbridge, "Global Climate Change," p. 431. Also Harrison and Prospero, "Reversals of the Earth's Magnetic Field," pp. 563–564.

62   *Effect of reversals on living things*: as reported by Hays and Opdyke, "Antarctic Radiolaria, Magnetic Reversals, and Climatic Change," pp. 1001–1011. Also Wyllie, *The Way the Earth Works*, p. 225.

63   *Scientists question theory of extinctions caused by reversals*: see, for example, Plotnick, "Relationship Between Biological Extinctions and Geomagnetic Reversals."

63   *Response of living things to magnetic field*: as reported by Palmer, "Geomagnetism and Animal Orientation," pp. 54–57.

63   *Birds' use of magnetism for navigation*: as reported by Gould, "The Case for Magnetic Sensitivity in Birds and Bees," pp. 256–267. Also Zimmerman, "Probing the Mysteries of How Birds Can Navigate the Skies," pp. 52–61.

63   *Effect of geomagnetism on human beings*: as reported by Friedman and Becker, "Geomagnetic Parameters and Psychiatric Hospital Admissions," p. 626.

Chapter 4

68   *History of continental drift theory*: as described by Tarling and Tarling, *Continental Drift*, pp. 13–17.

69   *Land bridges and the fossil record*: as reported by Hallam, "Continental Drift and the Fossil Record," pp. 57–66.

70–72   *Wegener's life*: as reported by Bullen, "Alfred Lothar Wegener," pp. 214–217. Also Wyllie, *The Way the Earth Works*, pp. 22–24.

78   *Rifting in Afar Triangle*: as reported by Ruegg et al., "Geodesic Measurements," p. 817.

79–80   *Red Sea*: as described by Ross et al., "Red Sea Drillings," pp. 377–380; and Ross, "The Red Sea: An Ocean in the Making," pp. 75–77.

80–81   *Early hominids in Afar Triangle*: as described by Johanson and Edey, "Lucy," pp. 4–9. Quotation p. 8. See also Johanson and Edey, *Lucy: The Beginnings of Humankind*.

81–82   *Lake Baikal geology*: as described by Shipler, "Siberian Lake Now a Model of Soviet Pollution Control," p. 10. Also Chelminski, "Baikal Survives as a Prize for the Whole Planet," pp. 44–51.

82   *Lake Baikal endemic species*: as reported by Chelminski, "Baikal Survives as a Prize for the Whole Planet," pp. 49–51.

88   *Hypothesis of the shrinking earth*: see, for example, Uyeda, *The New View of the Earth*, p. 20.

88   *Theory of expanding earth*: as presented by Carey, *The Expanding Earth*, 1976.

89   *Quasar geodetic measurements*: as described by Shapiro et al., "Transcontinental Baselines and the Rotation of the Earth Measured by Radio Interferometry," pp. 920–922.

89–90   *Satellite geodetic study*: personal communication, David E. Smith, Goddard Space Flight Center, Greenbelt, Maryland, June 11, 1981. Also *Lageos Technical Bulletin*, vol. 1, no. 3 (April 1981), National Aeronautics and Space Administration, Goddard Space Flight Center.

Chapter 5

page
94   *Empedocles and his sandals*: legend as related by Durant, *The Life of Greece*, p. 357.
94   *Irish monks' description*: from Thorarinsson, *Hekla on Fire*, p. 6, as cited by Tazieff, *Craters of Fire*, p. 211.
94   *Reaction to eruption of Mount St. Helens*: as reported in *Newsweek*, June 2, 1980, p. 26.
95–96 *Eruption of Paricutín*: as reported by Bullard, *Volcanoes of the Earth*, pp. 272–279. Also Green, "Paricutín, the Cornfield That Grew a Volcano," pp. 130–156.
96   *Classification of volcanic eruptions*: as described by Francis, *Volcanoes*, pp. 108–115.
97–99 *Letter of Pliny the Younger*: letter to Cornelius Tacitus, *Pliny Letters and Panegyricus*, pp. 425–433.
99–101 *Eruption of Mount Pelée*: as described by Francis, *Volcanoes*, pp. 81–93.
101   *Ignimbrite eruptions*: as described by Francis, *Volcanoes*, pp. 202–214.
101–102 *Griggs report on Mount Katmai and quotation*: from Griggs, *The Valley of Ten Thousand Smokes*, pp. 1–2.
102   *Ignimbrite eruption near Naples*: as reported by Francis, *Volcanoes*, p. 119.
102   *Volcanoes in or near the sea*: as described by Francis, *Volcanoes*, p. 14.
103   *Volcanoes in the Pacific*: as reported by Wyllie, *The Way the Earth Works*, p. 84.
103–105 *Volcanic eruption in Mediterranean*: as described by Bullard, *Volcanoes of the Earth*, pp. 308–310.
105   *Discovery of guyots*: from Hess, "Drowned Ancient Islands of the Pacific Basin," pp. 772–775.
107   *Eruption on Taal*: as reported by Francis, *Volcanoes*, pp. 304–306.
108   *Electrical storms during eruption of Mount St. Helens*: as reported in *Newsweek*, August 3, 1981, p. 56.

Chapter 6

111   *Precursor earthquake signs at Haicheng*: as reported by Fung-Ming, "An Outline of Prediction and Forecast of Taicheng Earthquake of M = 7.3," pp. 11–19. Also *Newsweek*, June 16, 1975, p. 45; and *Time*, September 1, 1975, pp. 36–41.
113   *Precursor signs in Liaoning Province*: as reported by Fung-Ming, "An Outline of Prediction," p. 18. Also *Time*, September 1, 1975, p. 39.
113   *Flickering lights and booming sounds*: as described by Shapley, "Chinese Earthquakes: the Maoist Approach to Seismology," p. 657. Also Meredith, "Earthquake Research and Political Tremors in China," p. 14.
113   *Announcement at Kuan-t'un Commune*: as cited in *Earthquake Frontiers*, a Chinese publication quoted by *Time*, September 1, 1975, p. 41.
113   *Results of successful prediction at Haicheng*: as reported by Press, "Earthquake Prediction," pp. 14–23. Also in *Time*, September 1, 1975, pp. 36–41.
114–115 *Earthquake prediction in China*: as reported by Chi-Yuan, "General Conditions of Earthquake Studies and Actions in China," pp. 5–11.

page

115   *Discoveries in Soviet Union*: as described by Press, "Earthquake Prediction," p. 18.

116   *Dilation theory*: as explained by Press, "Earthquake Predictions," pp. 17–18. Also Scholtz, "Toward Infallible Earthquake Prediction," pp. 56–58.

116   *U.S. scientists fail to confirm consistently USSR results*: personal communication with Hiroo Kanamori, Seismological Laboratory, California Institute of Technology, September 1982.

117   *Organization of Chinese population for prediction*: as reported by Press, "Earthquake Predictions," p. 18.

117   *Precursor signs at Tangshan*: as reported in *Newsweek*, August 9, 1976, pp. 30–32.

117   *Tangshan earthquake*: as reported in *Time*, August 9, 1976, p. 23. Also *Newsweek*, August 9, 1976, pp. 30–32.

118   *Oaxaca prediction*: as reported by Kerr, "Earthquake Prediction: Mexican Quake Shows One Way to Look for the Big Ones," pp. 860–862.

120   *New Madrid earthquakes*: as described by Penick, "I Will Stamp on the Ground," pp. 83–84, quotations cited from eyewitnesses Jared Brooks and Dr. Daniel Drake. Also Agar et al., *Geology from Original Sources*, p. 360.

120   *Massive fault discovered*: as reported by Zoback et al., "Recurrent Intraplate Tectonism in the New Madrid Seismic Zone," pp. 971–975.

121   *High-risk zones in U.S.*: as reported by Press, "Earthquake Prediction," p. 14.

121   *Palmdale Pimple*: as described by Hammond, "Earthquakes: An Evacuation in China, a Warning in California," p. 538. Also Shapley, "Earthquakes: Los Angeles Prediction," pp. 535–537.

123   *Reconstruction of fall of Minoan civilization*: scenario proposed by Vitaliano, *Legends of the Earth*, pp. 209–217.

126   *Quotation from Heritage papyrus*: as cited by Vitaliano, *Legends of the Earth*, p. 253.

127   *Quotation from Ipuwer papyrus*: as cited by Vitaliano, *Legends of the Earth*, p. 253.

127   *Similarities of earthquake effects to Exodus story have been suggested*: see, for example, Bennett, "Geo-physics and Human History," pp. 127–156.

129   *Crossing of the Red Sea*: as interpreted by Vitaliano, *Legends of the Earth*, pp. 260–264.

129   *Timing of the Exodus*: as interpreted by Professor Hans Goedicke, Johns Hopkins University, reported by Wilford, "Professor Moves Back Date of Exodus to Israel," p. 1.

129–130   *Hatshepsut inscription quoted*: translated by Hans Goedicke, as cited by Wilford, "Professor Moves Back Date of Exodus to Israel," p. 2. Also personal communication with Hans Goedicke, September 1982.

Chapter 7

134   *Formation of raindrops*: as described by Chandler, *The Air Around Us*, p. 90.

136   *Size of Antarctic ice sheet*: as reported by Strahler, *Principles of Physical Geology*, p. 323.

136   *First Byrd quotation*: from Byrd, *Little America*, p. 78.

Notes

page

136–137   *Second Byrd quotation*: from Byrd, *Little America*, pp. 353–354.

138   *Glacial drift and loess*: as described by Strahler, *Principles of Physical Geology*, p. 356.

138–139   *Evidence of ice age in Sahara*: as reported by Fairbridge, "The Sahara Desert Ice Cap," pp. 66–73.

141   *Precambrian glaciation*: as described by Dott and Batten, *Evolution of the Earth*, pp. 193–194.

142   *Timing of recent ice ages*: as reported by Toon and Pollack, "Volcanoes and the Climate," p. 26.

142–143   *Correlation between sunspot cycle and weather*: as reported by Schneider and Mass, "Volcanic Dust, Sunspots, and Temperature Trends," pp. 741–742.

143   *Anomalies in solar activity and periods of unusual climate*: as interpreted by Eddy, "Maunder Minimum," pp. 1189–1201.

143   *Long-term changes in luminosity of the sun*: as calculated by Sagan and Mullen, "Earth and Mars: Evolution of Atmospheres and Surface Temperatures," pp. 52–56.

144   *Early atmosphere-created greenhouse effect*: as theorized by Sagan and Mullen, "Earth and Mars," pp. 52–56.

145   *Franklin quotation*: as cited by Toon and Pollack, "Volcanoes and the Climate," p. 26.

145   *Decrease of solar radiation at Mauna Loa*: as reported by Toon and Pollack, "Volcanoes and the Climate," pp. 8–9.

145–146   *El Chichón cloud*: effects reported by *Chicago Tribune*, June 5, 1982, p. 1, citing interview with James Pollack, Ames Research Center, NASA.

146   *Volcanic dust in glaciers*: as reported by Toon and Pollack, "Volcanoes and the Climate," p. 26.

147   *Theory of supernova causing glaciations*: as set forth by Hunt, "Possible Climatic and Biological Impact of Nearby Supernovae," pp. 430–431.

148   *Theory of galactic cloud*: as set forth by McCrea, "Ice Ages and the Galaxy," pp. 607–608.

148–149   *Milankovitch theory*: as explained and supported by Hays et al., "Variations in the Earth's Orbit: Pacemaker of the Ice Ages," pp. 1121–1132. Also Dott and Batten, *Evolution of the Earth*, p. 440.

153   *Postulate that ice age accelerated human evolution*: as described by Jastrow, *Until the Sun Dies*, pp. 124–127.

Chapter 8

159   *Marine harvest from continental shelves*: as reported by Wolff, "Land Beneath the Sea," p. 34.

160   *Marco Polo quotation*: from *The Travels of Marco Polo*, pp. 258–259 and 334–335.

162   *Anchovy harvest*: as described by Wolff, "The Land Beneath the Sea," p. 38.

163   *Iron deposits*: as described by Friedman and Sanders, *Principles of Sedimentology*, pp. 183–184.

165   *Uses of bitumen by Sumerians*: as described by Williamson and Daum, *The American Petroleum Industry*, pp. 4–6.

166   *Oil in salt mines*: as described by Williamson and Daum, *The American Petroleum Industry*, pp. 14–17.

page
166   *American Medicinal Oil Company*: history described by Williamson and Daum, *The American Petroleum Industry*, pp. 17–18.
166–167   *Role of Samuel Kier and George Bissell*: as related by Williamson and Daum, *The American Petroleum Industry*, pp. 17–24.
169–170   *Mecca formation*: Zangerl and Richardson, "The Paleoecological History of Two Pennsylvanian Black Shales."

Chapter 9

174   *Eiseley quotation*: from Eiseley, *The Immense Journey*, p. 27.
178   *Alteration of opal*: as described by Sorrel, *Minerals of the World*, p. 210.
178   *Alteration of obsidian*: as described by Francis, *Volcanoes*.
178   *Volcanic glass altered to zeolites and feldspar*: as reported by Friedman and Sanders, *Principles of Sedimentology*, pp. 141–142.
179–180   *African formation*: as reported by Friedman and Sanders, *Principles of Sedimentology*, p. 142.
180   *Formation of red beds*: as described by Walker, "Formation of Red Beds in Modern and Ancient Deserts," pp. 353–368.
181   *Petrified Forest*: as described by Ash and May, "Petrified Forest, the Story Behind the Scenery."
182   *Mark Twain quotation*: from Mark Twain, *Life on the Mississippi*, 1875.

Chapter 10

187   *Magnetite and olivine beach sands*: as described by Friedman and Sanders, *Principles of Sedimentology*, p. 70.
187   *Ilmenite and rutile beaches*: as described by Carson, *The Edge of the Sea*, p. 111.
188   *River transport of sand*: as reported by Krinsley and Smalley, "Sand," p. 288.
188   *Recycling of ancient sand*, as outlined by Dott and Batten, *Evolution of the Earth*, p. 222.
189   *Similarity of sand grains*: as described by Krinsley and Smalley, "Sand," p. 286.
189   *Sorting by wind and water*: as described by Krinsley and Smalley, pp. 288–289.
189   *Desert dust storms*: as described by George, *In the Deserts of This Earth*: p. 18.
189   *Dust in Atlantic*: as reported by Prospero and Nees, "Dust Concentration in the Atmosphere of the Equatorial North Atlantic," p. 1196.
190   *Desert sandstorms*: as described by George, *In the Deserts of This Earth*, p. 18.
190–191   *George quotation*: from George, *In the Deserts of This Earth*, pp. 18–19.
191   *Dust storms on Mars*: as reported by Pollack, "Mars," pp. 108–109.
191   *Sand deserts on Mars*: as described by El-Baz, "Expanding Desert," p. 40.
192   *Great Kobuk dunes*: as described by Ricciuti, "Roving Sands of the Arctic," pp. 48–49.
193   *Thesiger quotation*: from Thesiger, *Arabian Sands*, pp. 144–145, 149.
193–194   *Bagnold quotation*: from Bagnold, *The Physics of Blown Sand and Desert Dunes*, p. 250.
194   *Singing sands*: as explained by El-Baz, "Expanding Desert," p. 39.

page

195   *Fossil dunes of Permian and Jurassic*: as described by Strahler, *Principles of Physical Geology*, p. 361.

195   *Desertification*: as reported by El-Baz, "Expanding Desert," pp. 33–40; also Eckholm, "Spreading Deserts: Livelihoods in Jeopardy," p. 2.

197   *Pitting of sand grains*: as described by Krinsley and Smalley, "Sand," pp. 290–291.

197   *Laplace quoted*: from Laplace, *Philosophical Essays on Probability*, as cited by Krinsley and Smalley, "Sand," p. 286.

199   *Muir quotation*: from Muir, *Gentle Wilderness: The Sierra Nevada*, p. 146.

Chapter 11

204   *Gems found in Sri Lanka*: as described by Cooray, *An Introduction to the Geology of Ceylon*, pp. 196–202; also Chernush, "Dazzling Jewels from Muddy Pits Enrich Sri Lanka," pp. 69–74.

207   *Aquamarine found in Minas Gerais*: as mentioned by Desautels, *The Mineral Kingdom*, p. 88.

207   *Synthetic gemstones*: process described by Desautels, *The Mineral Kingdom*, pp. 42–43.

208–209   *Marco Polo quotation*: from *The Travels of Marco Polo*, p. 399.

209   *Aristotle account*: as cited by Desautels, *The Mineral Kingdom*, pp. 76–77.

209   *Diamonds in Brazil*, as reported by Desautels, *The Mineral Kingdom*, p. 79.

209   *First diamonds found in Africa*, as described by Green, "Diamond Diggers in Namibia Sift Ocean Sands for Gemstones," p. 49.

210   *Diamond pipes at Kimberley*, as described by Desautels, *The Mineral Kingdom*, p. 321.

210–211   *Mining kimberlite formations*: as reported by Desautels, *The Mineral Kingdom*, pp. 77 and 133–134.

211–212   *Mining on Namibian coast*: as described by Green, "Diamond Diggers in Namibia," pp. 49–57.

213   *Concentration of gold in lithosphere*: as reported by Caldwell, "Gold," p. 479.

214   *Age of gold deposits*: as reported by Rose, "Gold: Mining and Metallurgy," p. 482.

214   *Legend of the Golden Fleece*: as related by Durant, *The Life of Greece*, p. 44.

214   *The riches of Croesus*: as mentioned by Rose, "Gold: Mining and Metallurgy," p. 482.

214–215   *Quantity of South American gold*: as reported by Caldwell, "Gold," p. 479.

215–217   *Eliot Lord quotation*: from *Comstock Mining and Miners*, U.S. Geological Survey, monograph 4, in Agar et al., eds., *Geology from Original Sources*, pp. 459–464.

218   *Hydrothermal solutions in Comstock lode*: as reported by Eliot Lord in Agar et al., eds., *Geology from Original Sources*, p. 469.

218   *Gold vug at Cripple Creek*: eyewitness account cited by Sprague, *Money Mountain*, p. 275.

219   *Sixteen trillion atoms an hour*: crystal formation as described by Desautels, *The Mineral Kingdom*, p. 49.

219   *Eiseley quotation*: Eiseley, *The Immense Journey*, p. 27.

Chapter 12

Chapter 13

Chapter 14

page

257   *Age of Hawaiian chain*: as reported by Strahler, *Principles of Physical Geology*, p. 185. Also Grigg, "Darwin Point: A Threshold for Atoll Formation."

257   *Spiders colonize new lands*: as described by Carson, *The Sea Around Us*, pp. 90–92.

258   *Endemic flora and fauna of Hawaii*: as described by Cox et al., *Biogeography*, pp. 104–108. Also Merlin, *Hawaiian Forest Plants* and the exhibits in the arboretum and botanical gardens of Waimea Falls Park.

258–259   *Coral species in Hawaii and projections for future development*: personal interview with R. W. Grigg, Hawaii Institute of Marine Biology, May 1982.

259   *Atoll formation at Eniwetok*: as reported by Chesher, *Living Corals*, p. 30.

260   *Oahu may become a coral atoll*: theory as described by R. W. Grigg, personal interview, May 1982.

260   *Pollution of Kaneohe Bay*: as reported by Johannes, "Life and Death of the Reef," pp. 48–53.

261   *Corrective measures in Kaneohe Bay*: as described by R. W. Grigg, personal interview, May 1982.

262   *Eiseley quotation*: Eiseley, *The Immense Journey*, pp. 11–12.

263   *Galaxy moved in direction of Virgo*, as reported by Waldrop, "A Flower in Virgo," p. 953.

265   *Thomas quotation*: from Thomas, *The Medusa and the Snail*, p. 15.

265   *Muir quotation*: from Muir, *Gentle Wilderness: The Sierra Nevada*, pp. 111 and 139.

Bibliography

Agar, William, Richard Foster Flint, and Chester R. Longwell, editors and compilers. *Geology from Original Sources*. New York: Henry Holt and Company, 1929.

Aristotle. *History of Animals*. Translated by D'Arcy Wentworth Thompson. In *Works of Aristotle*. Chicago: Encyclopaedia Britannica, Inc., 1952.

Ash, Sidney R., and David D. May. "Petrified Forest: The Story Behind the Scenery." Holbrook, AZ: Petrified Forest Museum Association, 1969.

Asimov, Isaac. *The Wellsprings of Life*. New York: The New American Library, 1960.

Bagnold, R. A. *The Physics of Blown Sand and Desert Dunes*. London: Methuen, 1941.

Ballard, Robert D. "Dive into the Great Rift." *National Geographic*, vol. 147 (May 1975), pp. 603–615.

———. "Window on Earth's Interior." *National Geographic*, vol. 150 (August 1976), pp. 228–249.

Barry, Roger G., John T. Andrews, and Mary A. Mahaffy. "Continental Ice Sheets: Conditions for Growth." *Science*, vol. 190 (December 5, 1975), pp. 979–981.

Bennett, J. G. "Geo-physics and Human History: New Light on Plato's Atlantis and the Exodus." *Systematics*, vol. 1 (1963), pp. 127–156.

Berlitz, Charles. *The Bermuda Triangle*. Garden City, NY: Doubleday and Company, Inc., 1974.

Bishop, Barry C. "How We Climbed Everest." *National Geographic*, vol. 124 (October 1963), pp. 477–509.

Bozorth, Richard Milton. "Magnetism." In *Encyclopaedia Britannica* (1956), vol. 14, pp. 637–667.

Briggs, John C. *Marine Zoogeography*. New York: McGraw-Hill Book Co., 1974.

Bullard, Fred M. *Volcanoes of the Earth*. Austin, TX: University of Texas Press, revised edition 1976.

Bullen, K. E. "Alfred Lothar Wegener." In *Dictionary of Scientific Biography*. Edited by Charles Coulson Gillispie. Vol. 14, pp. 214–217. New York: Charles Scribner's Sons, 1970.

Butler, Francis H., *et al*. "History of the Mariner's Compass." In *Encyclopaedia Britannica* (1956), vol. 6, pp. 168–171.

Byrd, Richard E. *Alone*. New York: G. P. Putnam's Sons, 1938.

————. *Little America.* New York: G. P. Putnam's Sons, 1930.

Calder, Nigel. "Head South with All Deliberate Speed: Ice May Return in a Few Thousand Years." *Smithsonian,* vol. 8 (January 1978), pp. 32–40.

Caldwell, William Elmer. "Gold." In *Encyclopaedia Britannica* (1956), vol. 10, pp. 479–481.

Carey, S. Warren. *The Expanding Earth.* Amsterdam: Elsevier Scientific Publishing Co., 1976.

Carson, Rachel. *The Edge of the Sea.* Boston: Houghton Mifflin Co., 1955.

————. *The Sea Around Us.* New York: Oxford University Press, 1951.

Casanova de Seingalt. *Memoires.* English translation by A. Machen, London, 1922.

Casas, Bartolomé de las. *Historia de las Indias.* In *Colección de Documentos,* edited by El Marqués de la Fuensanta del Valle and D. José Rayón. Vols. 62, 63, and 64. Madrid, 1875.

Chandler, T. J. *The Air Around Us.* Garden City, NY: The Natural History Press for the American Museum of Natural History, 1969.

Chelminski, Rudolph. "Baikal Survives as a Prize for the Whole Planet." *Smithsonian,* vol. 6 (November 1975), pp. 44–51.

Chernush, Akosh. "Dazzling Jewels from Muddy Pits Enrich Sri Lanka." *Smithsonian,* vol. 11 (June 1980), pp. 68–74.

Chesher, Richard, and Douglas Faulkner (photographer). *Living Corals.* New York: Clarkson N. Potter, Inc., 1979.

Chi-Yuan, Cha. "General Conditions of Earthquake Studies and Actions in China." In *Proceedings of the Lectures by the Seismological Delegation of the People's Republic of China, August 1, 1976.* Edited by Paul M. Muller; translated by M. Ohnuki, Kwok Maw Ong, and Chia-Chun Chao. Pasadena, CA: Jet Propulsion Laboratory, pp. 5–11.

Climap Project members. "The Surface of the Ice-Age Earth." *Science,* vol. 191 (March 19, 1976), pp. 1131–1144.

Cooke, Robert. "Ocean of Discoveries Excites Scientists on Seabed," *Chicago Sun-Times* (reprinted from *Boston Globe*), March 26, 1982.

Cooray, P. G. *An Introduction to the Geology of Ceylon.* Colombo: National Museums of Ceylon Publication, 1967.

Corliss, John B., and Robert D. Ballard. "Oases of Life in the Cold Abyss." *National Geographic,* vol. 152 (October 1977), pp. 441–453.

Cox, Allan, editor. *Plate Tectonics and Geomagnetic Reversals.* San Francisco: W. H. Freeman and Company, 1973.

Cox, Allan, G. Brent Dalrymple, and Richard R. Doell. "Reversals of the Earth's Magnetic Field." *Scientific American,* vol. 216 (February 1967), pp. 44–54.

Cox, C. Barry, Ian N. Healey, and Peter D. Moore. *Biogeography.* New York: John Wiley and Sons, 1973.

Darwin, Charles. *The Voyage of the Beagle.* In *The Harvard Classics,* edited by Charles W. Eliot. New York: P. F. Collier and Sons Corporation, 1937.

Degens, E. T., et al. "Template Catalysis: Asymmetric Polymerization of Amino Acids on Clay." *Nature,* vol. 227 (August 1, 1970), p. 492.

Dennison, Brian, and V. N. Mansfield. "Recent Glaciations and Dense Interstellar Clouds." NAIC 55, National Astronomy and Ionosphere Center, Cornell University.

Desautels, Paul E. *The Mineral Kingdom.* New York: Madison Square Press, 1968.

Doake, Christopher. "A Possible Effect of Ice Ages on the Earth's Magnetic Field." *Nature,* vol. 267 (June 2, 1977), pp. 415–416.

Dott, Robert H., Jr., and Roger L. Batten. *Evolution of the Earth.* New York: McGraw-Hill Book Company, 1976.

Durant, Will. *The Life of Greece.* Vol. 2, *The Story of Civilization.* New York: Simon and Schuster, 1939.

Dyhrenfurth, Norman G. "Six to the Summit." *National Geographic*, vol. 124 (October 1963), pp. 460–477.

Earle, Sylvia A., and Al Giddings (photographer). *Exploring the Deep Frontier.* Washington, DC: National Geographic Society, 1980.

Eckholm, Erik. "Spreading Deserts: Livelihoods in Jeopardy." *Not Man Apart* (March 1978), pp. 2–4. San Francisco: Friends of the Earth.

Eddy, John. "Maunder Minimum." *Science*, vol. 192 (June 18, 1976), pp. 1189–1201.

Eiseley, Loren. *The Immense Journey.* New York: Random House, 1957.

El-Baz, Farouk. "Expanding Desert Creates Grim Beauty but Also Threatens Crucial Cropland." *Smithsonian*, vol. 8 (June 1977), pp. 33–41.

Elsasser, Walter M. "The Earth as a Dynamo." *Scientific American*, vol. 198 (April 1958), pp. 44–48.

Emiliani, Cesare, et al. "Paleoclimatological Analysis of Late Quaternary Cores from the Northeastern Gulf of Mexico." *Science*, vol. 189 (September 26, 1975), pp. 1083–1088.

Encyclopaedia Britannica. William Benton, publisher. Chicago: Encyclopaedia Britannica, Inc., 1956.

Fairbridge, Rhodes W. "Global Climate Change During the 13,500 B.P. Gothenburg Geomagnetic Excursion." *Nature*, vol. 265 (February 3, 1977), pp. 430–431.

———. "The Sahara Desert Ice Cap." *Natural History*, vol. 80, No. 5 (June–July 1971), pp. 66–73.

Francis, Peter. *Volcanoes.* London: Penguin Books Ltd., 1976.

Freuchen, Peter. *Peter Freuchen's Book of Arctic Exploration.* New York: Coward-McCann, Inc., 1962.

Friedman, Gerald, and John E. Sanders. *Principles of Sedimentology.* New York: John Wiley and Sons, 1978.

Friedman, H., and R. O. Becker. "Geomagnetic Parameters and Psychiatric Hospital Admissions." *Nature*, vol. 200 (1963), p. 626.

Fung-Ming, Chu. "An Outline of Prediction and Forecast of Haicheng Earthquake of M = 7.3." In *Proceesdings of the Lectures by the Seismological Delegation of the People's Republic of China, August 1, 1976.* Edited by Paul M. Muller; translated by M. Ohnuki, Kwok Maw Ong, and Chia-Chun Chao. Pasadena, CA: Jet Propulsion Laboratory.

Ganapathy, R. "A Major Meteorite Impact on the Earth 65 Million Years Ago: Evidence from the Cretaceous-Tertiary Boundary Clay." *Science*, vol. 209 (August 22, 1980), pp. 921–922.

Gansser, Augusto. *Geology of the Himalayas.* London: Interscience Publishers, a division of John Wiley and Sons, Ltd., 1964.

George, Uwe. *In the Deserts of This Earth.* Translated from the German by Richard and Clara Winston. New York: Harcourt Brace Jovanovich, 1977.

Gilbert, William. *De Magnete Magneticisque Corporibus et de Magno Magnete Tellure, Physiologia Nova Plurimis et Argumentis et Experimentis Demonstrata.* London, 1600. English translation by Silvanus P. Thompson. Edited by the William Gilbert Society. London: Chiswick Press, 1900.

Giles, H. A. *History of Chinese Literature.* New York: Appleton, 1928.

Goedicke, Hans. *Egypt and the Early History of Israel.* Baltimore: Johns Hopkins University Press, in press.

Gould, James L. "The Case for Magnetic Sensitivity in Birds and Bees." *American Scientist*, vol. 68, no. 3 (May–June 1980), pp. 256–267.

Green, James A. "Paricutín, the Cornfield That Grew a Volcano." *National Geographic*, vol. 85 (February 1944), pp. 130–156.

Green, Timothy S. "Diamond Diggers in Namibia Sift Ocean Sands for Gemstones." *Smithsonian*, vol. 12 (May 1981), pp. 49–57.

Grigg, R. W. "Darwin Point: A Threshold for Atoll Formation." *Coral Reefs*, New York: Springer-Verlag, in press, 1982.

Griggs, Robert F. *The Valley of Ten Thousand Smokes*. Washington, DC: National Geographic Society, 1922.

Hallam, A. "Continental Drift and the Fossil Record." *Scientific American*, vol. 227 (November 1972), pp. 57–66.

Hammond, Allen L. "Earthquakes: An Evacuation in China, a Warning in California." *Science*, vol. 192 (May 7, 1976), pp. 538–539.

———. "Is the Sun an Inconstant Star?" *Science*, vol. 191 (March 19, 1976), pp. 1159–1160.

Harrison, C. G. A., and J. M. Prospero. "Reversals of the Earth's Magnetic Field and Climatic Changes." *Nature*, vol. 250 (August 16, 1974), pp. 563–564.

Hays, J. D., John Imbrie, and N. J. Shackleton. "Variations in the Earth's Orbit: Pacemaker of the Ice Ages." *Science*, vol. 194 (December 10, 1976), pp. 1121–1132.

Hays, J. D., and Neil D. Opdyke. "Antarctic Radiolaria, Magnetic Reversals, and Climatic Change." *Science*, vol. 158 (November 24, 1967), pp. 1001–1011.

Heezen, Bruce C., and Ian D. MacGregor. "Riddles Chalked on the Ocean Floor." *Saturday Review*, February 19, 1972, pp. 55–61.

Heirtzler, James R., et al. "Marine Magnetic Anomalies, Geomagnetic Field Reversals, and Motions of the Ocean Floor and Continents." In *Plate Tectonics and Geomagnetic Reversals*, edited by Allan Cox, pp. 265–282.

——— and Emory Kristof (photographer). "Project FAMOUS: Where the Earth Turns Inside Out." *National Geographic*, vol. 147 (May 1975), pp. 586–603.

Herodotus. *The History of Herodotus*. Translated by George Rawlinson. Chicago: Encyclopaedia Britannica, Inc., 1952.

Hess, H. H. "Drowned Ancient Islands of the Pacific Basin." *American Journal of Science*, vol. 244 (November 1946), pp. 772–791.

Hooke, Robert. *Micrographia*. First published by the Royal Society, London, 1665. Dover edition, London: Constable and Company, Ltd., 1961.

Hopson, Janet L. "Miners Are Reaching for Metal Riches on the Ocean's Floor." *Smithsonian*, vol. 12 (April 1981), pp. 50–59.

Hornbein, Thomas F., and William F. Unsoeld. "The First Traverse." *National Geographic*, vol. 124 (October 1963), pp. 509–514.

Hunt, Garry E. "Possible Climatic and Biological Impact of Nearby Supernovae." *Nature*, vol. 271 (February 2, 1978), pp. 430–431.

Jastrow, Robert. *Until the Sun Dies*. New York: W. W. Norton and Company, Inc., 1977.

Johannes, R. E. "Life and Death of the Reef." *Audubon*, vol. 78 (September 1976), pp. 38–55.

Johanson, Donald, and Maitland Edey. "Lucy." *University of Chicago Magazine*, vol. 73, no. 4 (Spring 1981), pp. 4–9.

———. *Lucy: The Beginnings of Humankind*. New York: Simon and Schuster, 1981.

Kerr, Richard A. "Earthquake Prediction: Mexican Quake Shows One Way to Look for the Big Ones." *Science*, vol. 203 (March 2, 1979), pp. 860–862.

Krinsley, D. H., and I. J. Smalley. "Sand." *American Scientist*, vol. 60 (May-June 1972), pp. 286–291.

Kukla, George J. "Around the Ice Age World." *Natural History*, vol. 85, no. 4 (April 1976), pp. 56–61.

Laplace, Pierre Simon de. *Philosophical Essays on Probability.* New York: John Wiley and Sons, 1956.

Latil, Pierre de, and Jean Rivoire. *Man and the Underwater World.* Translated by Edward Fitzgerald. London: Jarrolds Publishers Ltd., 1954, 1956.

Leonardo da Vinci. *The Notebooks of Leonardo da Vinci.* Arranged and translated by Edward MacCurdy. New York: Reynal and Hitchcock, 1938.

Lucretius. *De Rerum Natura.* Translated by T. Jackson. New York: Oxford University Press, 1929.

Lyell, Charles. *Principles of Geology.* First edition, London: J. Murray, 1830–1833.

McCray, Richard. "Molecules Between the Stars." *Natural History,* vol. 83, no. 10 (December 1974), pp. 72–77.

McCrea, W. H. "Ice Ages and the Galaxy." *Nature,* vol. 255 (June 19, 1975), pp. 607–608.

McDowell, Bart. "Avalanche! 3,500 Peruvians Perish in Seven Minutes." *National Geographic,* vol. 121 (June 1962), pp. 855–880.

———. "Earthquake in Guatemala." *National Geographic,* vol. 149 (June 1976), pp. 810–829.

Madariaga, Salvador de. *Christopher Columbus: Being the Life of the Very Magnificent Lord Don Cristóbal Colón.* New York: Macmillan Company, 1940.

Marden, Luis. "Man's New Frontier." *National Geographic,* vol. 153, no. 4 (April 1978), pp. 495–531.

Matthiessen, Peter. *The Snow Leopard.* New York: The Viking Press, 1978.

Merlin, Mark David. *Hawaiian Forest Plants.* Honolulu: Oriental Publishing Company, 1976.

Meredith, Dennis. "Earthquake Research and Political Tremors in China." *Technology Review* (October 1978), pp. 15–18.

Miers, Henry Alexander. "Diamond." In *Encyclopaedia Britannica* (1956), vol. 7, pp. 315–320.

Morison, Samuel Eliot. *Admiral of the Ocean Sea: A Life of Christopher Columbus.* Boston: Little, Brown and Company, 1942.

Morison, Samuel Eliot, editor and translator. *Journals and Other Documents on the Life and Voyages of Christopher Columbus.* New York: The Heritage Press, 1963.

Muir, John. *Gentle Wilderness: The Sierra Nevada.* Edited by David Brower, Sierra Club. New York: Ballantine Books, 1967.

Nicolson, Marjorie Hope. *Mountain Gloom and Mountain Glory.* New York: W. W. Norton and Company, 1959.

Paecht-Horowitz, M., J. Berger, and A. Katchalsky. "Prebiotic Synthesis of Polypeptides by Heterogeneous Polycondensation of Amino-acid Adenylates," *Nature,* vol. 228 (November 14, 1970), pp. 636–639.

Palmer, John D., "Geomagnetism and Animal Orientation." *Natural History,* vol. 76 (November 1967), pp. 54–57.

Parasnis, D. S. *Magnetism from Lodestone to Polar Wandering.* New York: Harper and Brothers, 1961.

Pausanias, *Description of Greece.* Translated by W. H. S. Jones. The Loeb Classical Library. Vols. 1–3, New York: G. P. Putnam's Sons; vols. 4, 5, Cambridge, MA: Harvard University Press, 1918–1935.

Penick, James, Jr. "I Will Stamp on the Ground with my Foot and Shake Down Every House." *American Heritage,* vol. 27 (December 1975), pp. 82–87.

Perkins, Dexter, Jr. "Ice Age Animals of the Lascaux Cave," *Natural History,* vol. 85, no. 4 (April 1976), pp. 62–69.

Pliny the Younger. *Pliny Letters and Panegyricus,* vol. 1. Translated by Betty

Radice. The Loeb Classical Library. Cambridge, MA: Harvard University Press, 1969.

Plotnick, Roy. "Relationship Between Biological Extinctions and Geomagnetic Reversals." *Geology*, vol. 8 (December 1980), pp. 578–581.

Pollack, James B. "Mars." *Scientific American*, vol. 233 (September 1975), pp. 106–117.

Polo, Marco. *The Travels of Marco Polo*. Revised from Marsden's translation and edited by Manuel Komroff. New York: The Heritage Press, 1934.

Porter, Roy. *The Making of Geology: Earth Science in Britain 1660–1815*. Cambridge: Cambridge University Press, 1977.

Press, Frank. "Earthquake Prediction." *Scientific American*, vol. 232 (May 1975), pp. 14–23.

Prospero, Joseph M., and Ruby T. Nees. "Dust Concentration in the Atmosphere of the Equatorial North Atlantic: Possible Relationship to the Sahelian Drought." *Science*, vol. 196 (June 10, 1977), pp. 1196–1198.

Ratcliffe, J. A. *Sun, Earth, and Radio*. New York: McGraw-Hill Book Company for World University Library, 1970.

The Rand McNally Atlas of Oceans. Edited by Martyn Bramwell. New York and Chicago: Rand McNally and Company, 1977.

Ricciuti, Edward R. "Roving Sands of the Arctic." *Audubon*, vol. 81 (March 1979), pp. 44–53.

Roessler, Carl. *The Underwater Wilderness: Life Around the Great Reefs*. New York: Chanticleer Press, 1977.

Rogers, Eric M. *Physics for the Inquiring Mind*. Princeton, NJ: Princeton University Press, 1960.

Rose, Thomas Kirke. "Gold: Mining and Metallurgy." In *Encyclopaedia Britannica* (1956), vol. 10, pp. 481–484.

Ross, David A. "The Red Sea: An Ocean in the Making." *Natural History*, vol. 85, no. 6 (August–September 1976), pp. 75–77.

—— et al. "Red Sea Drillings." *Science*, vol. 179 (January 26, 1973), pp. 377–380.

Rudwick, Martin. *The Meaning of Fossils*. New York: Science History Publications, USA, a division of Neale Watson Academic Publications, Inc., revised edition 1976.

Ruegg, J. C., et al. "Geodetic Measurements of Rifting Associated with a Seismovolcanic Crisis in Afar." *Geophysical Research Letters*, vol. 6, no. 11 (November 1979), pp. 817–820.

Sagan, Carl. "Beginnings and Ends of the Earth." *Natural History*, vol. 82 (October 1973), p. 101.

—— and George Mullen. "Earth and Mars: Evolution of Atmospheres and Surface Temperatures." *Science*, vol. 177 (July 7, 1972), pp. 52–56.

Schneider, Stephen H., and Clifford Mass. "Volcanic Dust, Sunspots, and Temperature Trends." *Science*, vol. 190 (November 21, 1975), pp. 741–742.

Scholtz, Christopher H. "Toward Infallible Earthquake Prediction." *Natural History*, vol. 83, no. 5 (May 1974), pp. 54–59.

Schopf, Thomas J. M. *Paleoceanography*. Cambridge, MA: Harvard University Press, 1980.

Shapiro, I. I., et al. "Transcontinental Baselines and the Rotation of the Earth Measured by Radio Interferometry." *Science*, vol. 186 (December 6, 1974), pp. 920–922.

Shapley, Deborah. "Chinese Earthquakes: the Maoist Approach to Seismology." *Science*, vol. 193 (August 20, 1976), pp. 656–657.

——. "Earthquakes: Los Angeles Prediction Suggests Faults in Federal Policy." *Science*, vol. 192 (May 7, 1976), pp. 535–537.

Sheppard, Asher R., and Merril Eisenbud. *Biological Effects of Electric and Magnetic Fields of Extremely Low Frequency.* New York: New York University Press, 1977.

Shipler, David K. "Siberian Lake Now a Model of Soviet Pollution Control." *The New York Times*, April 16, 1978.

Sisson, Robert F. "Life Cycle of a Coral." *National Geographic*, vol. 143 (June 1973), pp. 780–793.

Sorrel, Charles A. *Minerals of the World.* New York: Golden Press, 1973.

Sprague, Marshall. *Money Mountain: The Story of Cripple Creek Gold.* Boston: Little, Brown and Company, 1953.

Strahler, Arthur N. *Principles of Physical Geology.* New York: Harper and Row, 1977.

Sullivan, Walter. "Deep-Sea Life is Found Flourishing on Sulfur from Ocean's Volcanoes." *The New York Times*, April 25, 1982.

Tarling, M. P., and D. H. Tarling. *Continental Drift.* London: Penguin Books Ltd., 1972.

Tazieff, Haroun. *Craters of Fire.* New York: Harper and Brothers, 1952.

Thesiger, Wilfred. *Arabian Sands.* London: Penguin Books Ltd., 1964.

Thomas, Lewis. *Lives of a Cell.* New York: The Viking Press, Inc., 1974.

———. *The Medusa and the Snail.* New York: The Viking Press, Inc., 1979.

Thorarinsson, Sigurdur. *Hekla on Fire.* Munich: Hans Reich Verlag, 1956.

Toon, Owen B., and James B. Pollack. "Volcanoes and the Climate." *Natural History*, vol. 86, no. 2 (January 1977), pp. 8–26.

Uyeda, Seiya. *The New View of the Earth.* Translated by Masako Ohnuki. San Francisco: W. H. Freeman and Company, 1978.

Van Andel, Tjeerd. *Tales of an Old Ocean.* New York: W. W. Norton and Company, Inc., 1977.

Van Valen, Leigh, and Robert E. Sloan. "Ecology and the Extinction of the Dinosaurs." *Evolutionary Theory*, vol. 2 (1977), pp. 37–64.

Vine, Fred J., and Drummond H. Matthews. "Magnetic Anomalies over Oceanic Ridges." In *Plate Tectonics and Geomagnetic Reversals*, edited by Allan Cox, pp. 265–282.

Vitaliano, Dorothy B. *Legends of the Earth.* Bloomington, IN: Indiana University Press, 1973.

Waldrop, M. Mitchell. "A Flower in Virgo." *Science*, vol. 215 (February 19, 1982), pp. 953–955.

———. "Mount St. Helens: Researchers Sort the Data." *Chemical and Engineering News* (June 9, 1980), pp. 19–20.

———. "Ocean's Hot Springs Stir Scientific Excitement." *Chemical and Engineering News* (March 10, 1980), pp. 30–33.

Walker, T. B. "Formation of Red Beds in Modern and Ancient Deserts." *Geological Society of America Bulletin*, vol. 78 (1967), pp. 353–368.

White, Edmund. "The Chinese Landscape." *Horizon*, vol. 17, no. 4 (Autumn 1975), pp. 87–97.

Wilford, John Noble. "Professor Moves Back Date of Exodus to Israel," *The New York Times*, reprinted in *Arizona Daily Star*, May 4, 1981.

Williamson, Harold F., and Arnold R. Daum. *The American Petroleum Industry.* Evanston, IL: Northwestern University Press, 1950.

Wilson, Leonard G. *Charles Lyell, the Years to 1841, the Revolution in Geology.* New Haven.: Yale University Press, 1972.

Wolff, Anthony. "Land Beneath the Sea." *The Lamp*, vol. 59, no. 3 (Fall 1977), pp. 34–41. New York: Exxon Corporation.

Wyllie, Peter J. *The Way the Earth Works.* New York: John Wylie and Sons, Inc., 1976.

Zangerl, Rainer, and Eugene S. Richardson, Jr. "The Paleoecological History of Two Pennsylvania Black Shales." *Fieldiana: Geology Memoirs*, vol. 4 (April 1963), pp. 1–352.

Zimmerman, David R. "Probing the Mysteries of How Birds Can Navigate the Skies," *Smithsonian*, vol. 10 (June 1979), pp. 52–61.

Zoback, M. D. et al. "Recurrent Intraplate Tectonism in the New Madrid Seismic Zone." *Science*, vol. 209 (August 29, 1980), pp. 971–976.

Index

Page numbers in *italic* indicate the location of definitions of terms.

Index